CELESTION

———— 1924–2024 ————

A CENTURY OF SOUND

THE STORY OF CELESTION

CONTENTS

Foreword by Brian May 6

PART ONE, 1924-1949
EARLY BEGINNINGS, WAR AND ROLA 10

Preface: The Brave New World of Broadcasting 12

Chapter 1: The French Revolution 16

Chapter 2: The Very Soul of Music 24

Chapter 3: Recession, War and Rola 34

    The Rola Model G12 40

Chapter 4: Rola Celestion 42

PART TWO, 1950-1970
TRANSFORMATION AND EVOLUTION 48

Chapter 5: Truvox and the Empire of Daniel Prenn 50

Chapter 6: Vox, Marshall and the Evolution of the G12 58

Chapter 7: Big Sounds, Big Names 72

Chapter 8: HiFi Comes of Age 82

Chapter 9: Celestion Industries, plc 88

PART THREE, 1971-1991
POWER TO THE PEOPLE 92

Chapter 10: Two Decades of Turbulence 94
    The Prenn Dynasty 105

Chapter 11: The Golden Decades of HiFi 106
    How the Drinkmaster™ Capsule Became an Aluminium Dome 118

Chapter 12: G12 and the Quest for More Power 120
    Mysteries of the G12 Revealed 128

Chapter 13: Sound Reinforcement Systems 132

Chapter 14: Celestion International 142

PART FOUR, 1992-2024
TECHNOLOGY LEADS THE WAY 148

Chapter 15: The Creation of KH Manufacturing 150

Chapter 16: Celestion Professional 158

Chapter 17: New Beginnings 172

Chapter 18: The Voice of Rock & Roll 180
    Partners in Tone 193

Chapter 19: A Professional Audio Future 196

Appendix I: The T-Book 210

Appendix II: Patents Granted 216

Appendix III: Managing Directors/General Managers 222

Appendix IV: Celestion Date Codes 224

Acknowledgments 226

FOREWORD
A few words about Celestion from a friend...

August 1993, London.

My first encounter with a Celestion speaker was in 1962, in a rather flea-bitten second-hand equipment store in Wardour Street.

My first amplifier had been the family Radiogram which my amazing dad had built, along with almost every other piece of electrical equipment in our house. The sound it made hooked up to a coil of wire which I had wound around some Eclipse magnets and fixed to my Spanish guitar was exciting enough to convince me that this was the way to go! When I eventually saved up enough pennies to invest in an amplifier of my own, I was persuaded that I should go for something more 'state of the art', something using the new wonder of the age – transistors! Bad idea. The amplifier was loud, but did not have that magical metallic, yet smooth characteristic which I had already fallen in love with.

It was because I had briefly tried out a Vox AC30 belonging to a friend that I had made the trip to the second-hand shop. There I hoped I would find the sound I already had in my head. I went there with a couple of friends from the newly formed 'Smile' group, which was later to mysteriously metamorphose into another group called 'Queen'. The man hauled out two somewhat battered AC30s and I plugged in (via a small treble booster).

That was it! It was all there, without a moments doubt! "They're all original, hardly used, and look," he said, pointing to the back of the cabinets, "with the original Celestions!"

Quite why AC30s sound so perfect to people like myself is quite a mystery. No-one has ever satisfactorily explained it to me. But there again it doesn't matter since I have since learned that if it sounds right, it is right; a useful rule of thumb, since it tends to divert the musician's attention to where it should be, in the music itself.

But my father was in possession of many of the clues. He was an electronics engineer and had spent most of his life immersed in the theory and practice of equipment based on electronic tubes. To him KT66s, EL34s and multi-tap transformers were the stuff of everyday life, much as mobile phones, faxes and personal computers are to the young adults of today. So perhaps it is not so

Some of Brian's Vox AC30s, including a monochrome cabinet from 1964.

surprising that those valve amplifiers and those speakers have never been bettered to this day. Technology has moved to a different place, but far enough away to leave the culture or cultivation of tube amplifiers to an enthusiastic few.

One thing that has remained clear to me is that those Celestion Blues were a vital part of the sound. And look in the back of any guitarist's amplifier in 1993 and the chances are that you will find something very similar to those speakers every time.

Of course, a whole mystique grew up around such things along the way. Did they have to be blue? Was a particular vintage blessed with a perfect sound? Did they have to be mounted in twos or would they perform just as well in fours?

Who knows? But listen to the recordings of the Beatles, The Stones, Clapton, Hendrix, Satriani, Steve Vai… even that old 'Queen' group, and what you are listening to is the sound of one of those trusty Celestion G12s vibrating faithfully in tune with the string which vibrated under the guitarist's finger at that moment when a piece of history was made.

Cheers guys, and here's to the future…

August 2023, London.

Well dear folks, that was what I wrote 30 years ago about the legend of Celestion.

The amazing thing is that even though the whole world has vastly changed since then, the legend is still alive and well. I could rewrite this piece right now and change the mention of 'faxes' (do you remember them?) with the Internet and Instagram, and all my comments about the place of those 'Blues' in Rock Music would still hold good.

To the best of my knowledge there's no such thing as a digital loudspeaker. No matter what processes the guitar signal goes through in the digital domain, it involves a speaker that still physically moves the air in tune with its own vibrations. And the air moves the analogue apparatus in our heads to produce the sensation of sound.

For some reason, Celestion got it right first time; they still sound best. So, the Celestion speaker has firm and secure place in the future as well as the past. All I can say is that makes me strangely happy!

Rock on guys!

Bri.

1924–1949

1
Early Beginnings, War and Rola

High Street, Hampton Wick c. 1920.

PREFACE

The Brave New World of Broadcasting

The 1920s that Celestion was born into was a decade of contrasts. Post-war prosperity, and the arrival of the Roaring Twenties, fuelled by the US economic boom, had quickly extinguished memories of the deadly Spanish flu pandemic that had claimed the lives of 50 million across the globe, just as the First World War was ending.

The gloriously scandalous Jazz Age also ushered in the Golden Era of Radio once the Westinghouse Company had been granted a licence by the US Government. Following its first, historic broadcast on Pittsburgh's KDKA in November 1920, the company wasted no time in making radios available to the general public. This ensured that the twenties would roar for the stay-at-homes as much as for the flappers and those out strutting Charlestons and Black Bottoms.

By the end of 1922 there were over 500 licensed stations operating in the US; and over in the UK that same June, Britain's first live public broadcast was made from the factory of Marconi's Wireless Telegraph and Signal Company in Chelmsford, Essex, featuring celebrated Australian soprano Dame Nellie Melba. By November daily radio broadcasts had begun, with a 6pm news bulletin followed by a weather forecast prepared by the British Meteorological Office.

1922 was also the year the classic *Marconiphone* wireless equipment, an eccentric two-valve receiver, entered the market and was shown to the British public at the All-British Wireless Exhibition in London. The audience reaction triggered a tidal wave of applications for experimental broadcast licences, all hoping to provide the British listening public with an alternative programming source to their trusty phonographs with room-dominating horns.

Instead, the UK Government, via the General Post Office (GPO), issued just a single broadcasting licence. This was to a company owned by a British-American consortium of leading wireless receiver manufacturers known as the British Broadcasting Company Ltd. Fairly short-lived, it was the forerunner to the behemoth that would become the BBC.

Originally incorporated in October 1922, when the company was dissolved in 1926 its assets were passed over to the non-commercial organisation that we know today.

Opposite: Radio Times, Christmas 1923

THE RADIO TIMES

THE CHRISTMAS NUMBER

"JUST A SONG AT TWILIGHT"

6ᴅ

The BBC began radio broadcasting on 14 November 1922, on a wavelength of 369 metres with the call sign 2LO. After the success of the London station, two further stations were opened at Birmingham on 5IT and Manchester on 2ZY. The six shareholders agreed to give the BBC the benefit of their respective patents, and only radio sets supplied by BBC companies were permitted to be licensed to receive programmes. Within three years of the first complete wireless receivers appearing in the UK, some 85% of the population were able to pick up these transmissions.

How did this new sound source arrive in its listeners' ears? And what has any of it to do with Celestion? In 1923 crystal (or 'cat's whisker') sets were the most popular due to their low cost and freedom from expensive high-tension batteries. A single crystal set only had enough output to operate headphones, which comprised two coils wound on a horseshoe magnet vibrating a metal plate, although, at a pinch, it was possible to connect several of them to a receiver so that a family could listen together.

The more powerful valve (vacuum tube) radio receivers had the ability to operate the recently introduced horn loudspeakers, built in the shape of a free-standing horn reminiscent of the ornate specimens beloved by gramophone enthusiasts. However, these cumbersome appliances gradually gave way to the separate loudspeaker cabinet, often highly decorative and taking pride of place in the lounge or front room: a piece of furniture as much to be displayed as listened to. And with the increase in the availability of wireless receiving sets that appeared as a result of the BBC stepping up transmitter power, the need for loudspeakers increased exponentially.

In the early 1920s former gramophone maker Cyril French, alongside his three brothers, had already set up a small manufacturing business in the picturesque village of Hampton Wick to the south-west of London. This was his cue to set foot in the fledgling world of broadcast audio.

1924 saw French working feverishly on improving a design conceived by Eric Vincent Mackintosh for one of the earliest cone loudspeakers. It's unclear how the two men met, but the timing was perfect, as by the end of the year more than a million receiving licences had been issued by the GPO and the BBC had 20 radio-transmitting stations in operation.

So it was that 1924 became the year that gave birth to *Celestion*—initially as the name of a product produced by the French family firm, before becoming a company in its own right, with Cyril and Eric blazing a trail at its helm. In the world of British audio, Celestion predated Tannoy by two years and Vitavox by seven.

In many other respects 1924 became the landmark year for broadcasting. It was the year that Big Ben and the Greenwich Time Signal (affectionately known as 'the pips') rang across the airwaves for the first time (the six-pip time signal being introduced in February following the successful broadcast of the chimes of Big Ben to usher in the New Year). Television also made

Celestion founders Cyril French (*left*) and Eric Mackintosh (*centre*) and the premises at 29 High St., Hampton Wick (*right*).

its first appearance when in Hastings, East Sussex, John Logie Baird wirelessly transmitted rudimentary moving pictures over a short distance.

Listeners with access to this brave new world of broadcasting heard for the first time that enduring British radio phenomenon, the Shipping Forecast, the world's first radio play, and most significantly of all, the first royal broadcast by King George V, when he opened the grand colonial British Empire Exhibition at Wembley Stadium in April that same year.

The year which began with Ramsay MacDonald leading Britain's first Labour Government ended with Stanley Baldwin running the country for the Conservatives. Over in the USSR Lenin died, while in the United States Gershwin completed *Rhapsody in Blue* and Chrysler produced its first car. Closer to home, in the world of transducers, 1924 was also the year that the moving-coil principle was patented.

Thus, Mackintosh and French set off on their adventure in the era of the radiogram, writing the early chapters of this story. By laying down fundamental principles still recognisable to this day, they underpinned a future that would ensure both prosperity and, when necessary, survival through the choppiest of waters.

Celestion is proud to be the first transducer company in the loudspeaker industry to have reached the 100-year landmark having remained in continuous operation, still designing and unwaveringly delivering the same type of product now as it did then. Its skilled engineers down the generations have gone on to influence other areas of the audio industry, and in so doing, have left a trail of classic products in their wake. This is the story of how Celestion's remarkable journey has unfolded.

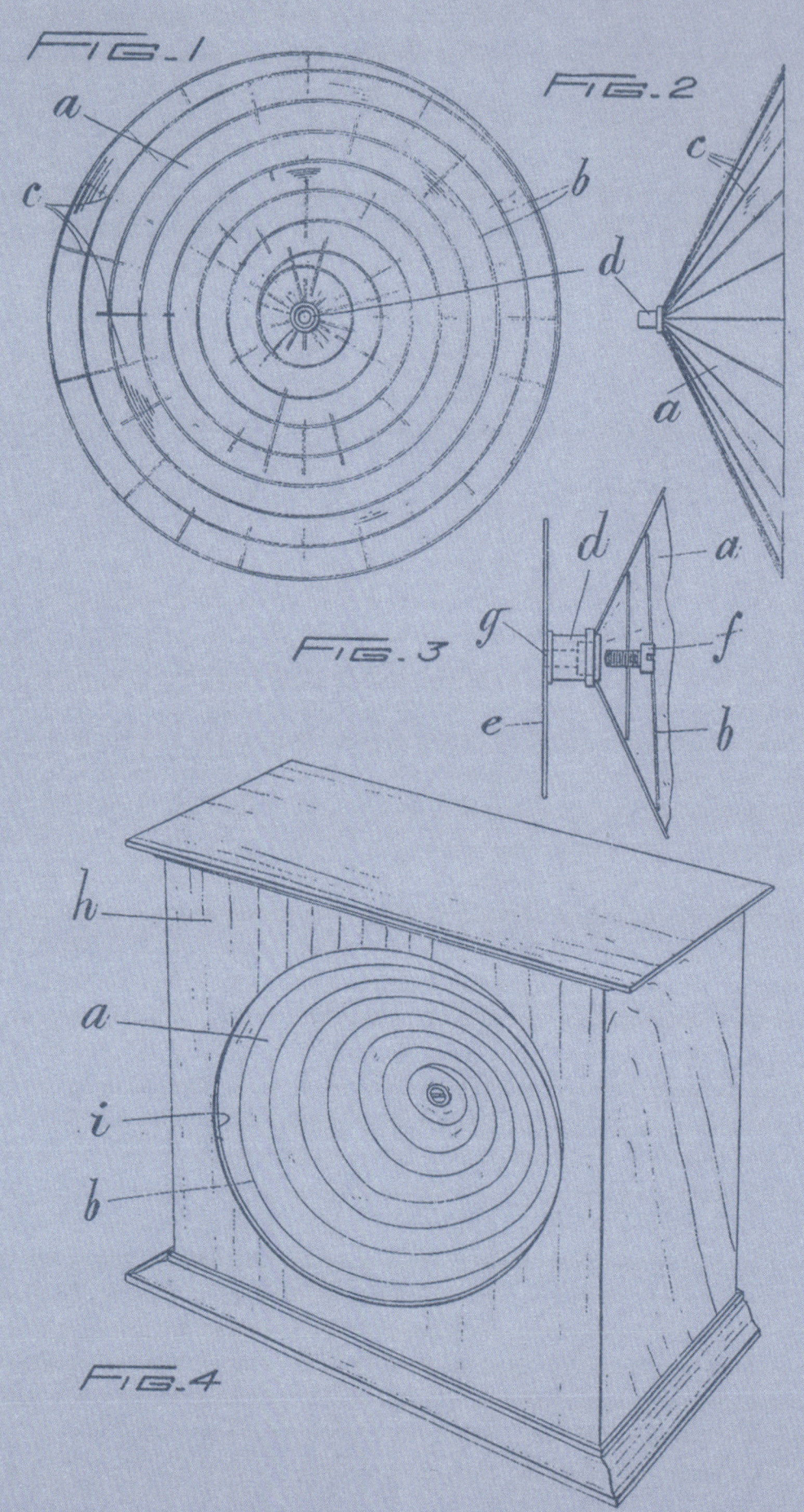

1 The French Revolution

It was the invention of a new kind of loudspeaker, conceived by Eric Mackintosh and finally made real by the practical expertise of Cyril French, which gave life to Celestion. The story of the origins of the company is, in truth, the story of these two men and how they came to meet in a small village located just to the south-west of London.

Towards the end of the nineteenth century, the French family lived in Whittlesford, Cambridgeshire, where father Herbert was proprietor of the Bridge Inn at nearby Waterbeach. Cyril, born in 1886, was the second of four sons; the eldest was Ralph, the younger two, Leonard and Edgar.

According to the 1911 census, Ralph was an engineer's draughtsman in Letchworth in the neighbouring county of Hertfordshire. Cyril was an instrument maker, producing 'water meters, etc.' Edgar was a chef in a hotel in Wales (some 200 miles away), while Leonard was a carpenter and joiner who remained at home.

The previous year, Cyril had completed a six-year apprenticeship at the Cambridge Scientific Instrument Co., a company that had emerged to supply the prestigious University of Cambridge with instrumentation. It was a company that could already boast such alumni as Horace Darwin (youngest son of botanist Charles) and William T. Pye, who later went on to form his eponymous, world-famous audio and electronics firm with son William G. Pye. Later Cyril worked with G. Kent & Co. and Walters Electrical Manufacturing (an early manufacturer of radio receivers), and by 1919 he had joined J.E. Jaccard, 'Manufacturer of Talking Machines' (i.e. early gramophones).

However, it was Herbert French and son Leonard who, by 1921, had made the move to South-West London. Leonard is listed in the 1921 census as being an Electro Plater, with an address of 85 High Street, Hampton Wick.

Opposite: Drawing from Eric Mackintosh's original patent application.

Clearly a talented inventor, by 1923 he had already applied for two separate patents relating to 'thermally activated switches'

Perhaps it was the change of career that precipitated Leonard and Herbert's move south but it must have been successful, as Cyril and Edgar soon joined them. The family later took over the two-storey Thames Valley Plating Works at 29 High Street, Hampton Wick, and soon after renamed it the Electrical Manufacturing and Plating Company. Its official listing was 'instrument manufacturers', and on his arrival, Cyril wasted no time getting to work. Clearly a talented inventor, by 1923 he had already applied for two separate patents relating to 'thermally activated switches'.

As for Eric Vincent Mackintosh, he was born in West Dulwich (London) on 26 June 1890— the youngest child of Alexander and Annie, who hailed originally from Inverness, Scotland. He joined the Royal Navy in November 1914; his occupation at the time was stated as 'musician' and he was reported to have had 'prodigious musical ability'.

During the First World War, German submarines became a particular problem for Britain and its allies, inflicting heavy losses on the naval and mercantile fleets. Urgent countermeasures were necessary, and among these was the rapid development of the hydrophone, an underwater device for detection and location of hostile vessels.

Following his demobilisation from the Royal Navy in 1916, reportedly due to poor health, it's believed that Mackintosh spent the latter years of the war in civilian service, at one of the hydrophone listening stations in Malta. Later he transferred to Admiralty Experimental Station (AES) Shandon near Glasgow. Here, Eric joined the Acoustics group, under the leadership of B.S. Smith, whose focus was primarily on research around the generation and reception of sound underwater, including tests on a 'Malta pattern liquid-filled microphone'.

AES Shandon was closed in January 1921, and much of the work, and many of the workers, transferred to the newly established Admiralty Research Laboratory in Teddington, next door to the National Physical Laboratory. Eric was one of those transferred, and by June 1921 he and his family were resident a couple of miles away from the lab, at 18 Lower Teddington Road in Hampton Wick. By the close of 1923, Eric had filed his first patent relating to a unique design for cone loudspeakers.

Left: Diagram of the moving armature of *The Celestion*.

Right: Eric Mackintosh's original patent application, completed 10th September 1924.

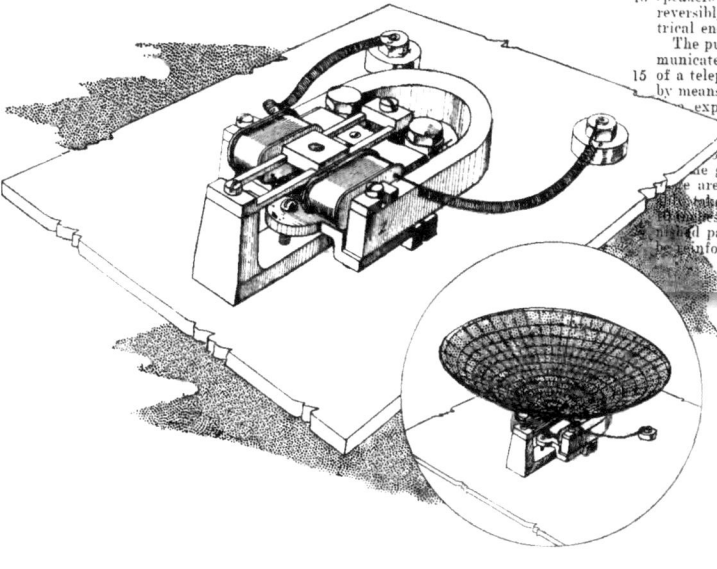

A little further down the High Street from the Plating Company lived J.C.C. Berry, Hon. Secretary of the *Kingston and District Radio Society*, whose regular meetings attracted radio enthusiasts from the local area. Could it have been at one of these get-togethers that these two audio aficionados finally met each other and began discussing their shared passion?

Amid what must have been great excitement, the two pioneers found each other and joined together to work on developing Eric's initial concept, finally producing a fully functioning loudspeaker design sometime in 1924. Joined by the oldest French brother, Ralph, Cyril and Eric began their loudspeaker manufacturing business at number 29 while Leonard and Edgar continued with the everyday operation of the plating business.

The original Mackintosh design was filed for patent on 16 December 1923, under the heading of 'Improvements in Telephone Receivers and Like Instruments'. The application was completed on 10 September 1924, and issued as British Patent No. 230,552 on 16 March 1925. It employed a free-vibrating outer edge, with clamping taking place at the central armature (which vibrated under the influence of an electrical signal). Prior to the introduction of permanent-magnet (PM) technology, the driving mechanism had been a moving-iron balanced-armature type.

THE FRENCH REVOLUTION

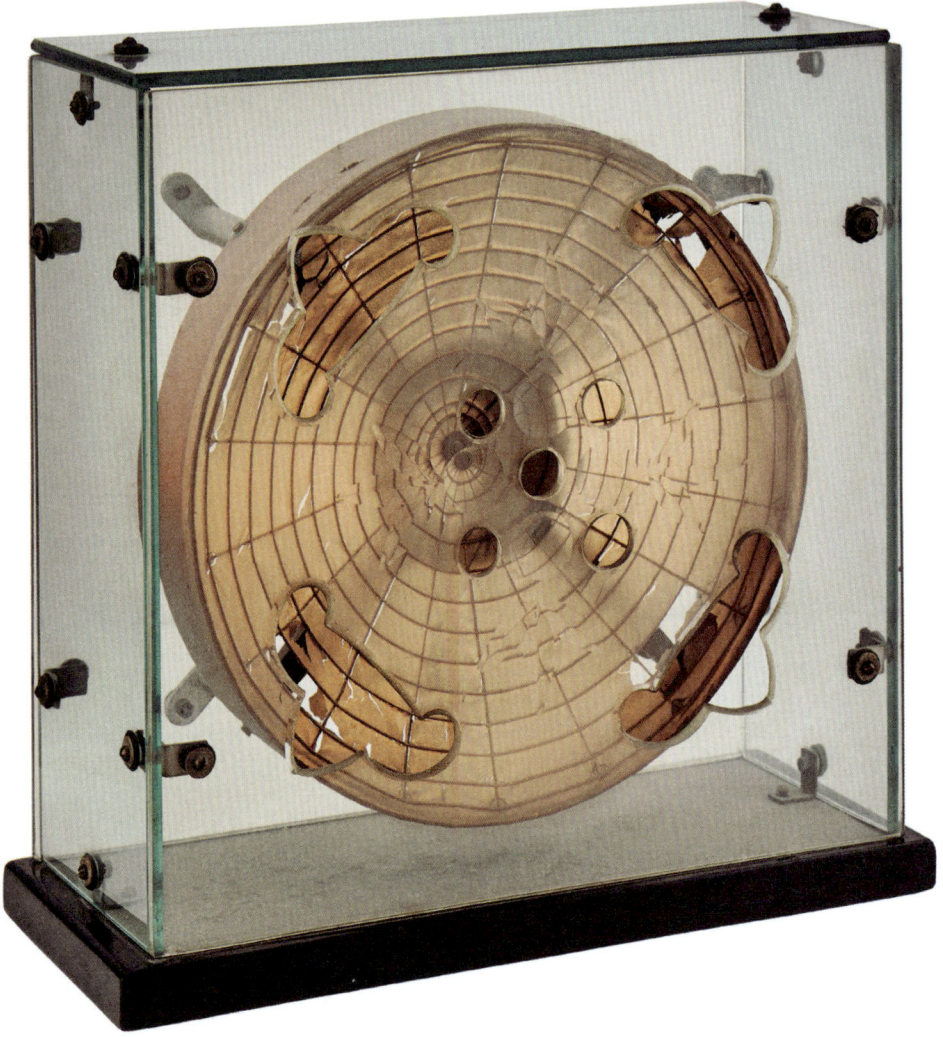

Glass display cabinet featuring the improved French/Mackintosh diaphragm with clamped outer edge.

In the patent description, Mackintosh stated that 'the purpose of the invention is to communicate the vibrations of the armature'—not so dissimilar to the way the loudspeakers of today work. The description continues:

> The diaphragm is mounted on and supported by the armature or vibrating member … It preferably takes the form of a cone of about 10in in diameter which may be of varnished paper or the like, and which may be reinforced by radial ribs and concentric rings or hoops of … material(s) suitable for use in a light and rigid framework.

So it was that the first housed loudspeaker of the Electrical Manufacturing and Plating Co. came to be: a transducer invented by Eric Mackintosh and realised by Cyril French, by all accounts turning a bright idea into a viable product. The name it was given was the rather ethereal and angelic-sounding 'Celestion', suggested in a moment of inspiration by Ralph French. It is hard to believe that something seemingly so delicate would emanate from a utilitarian-sounding plating company—the two things seemed worlds apart.

This instrument can be had in either oak, mahogany, or walnut, at prices ranging from £6.10s to £6.15s

When the Celestion loudspeaker was finally launched to the general public in January 1925, customers were given a choice of oak, walnut or mahogany for the fine wooden enclosure. The product was marketed with the playful strapline 'The Loud Speaker of Distinction', and with its delicate paper cone, reinforced with the thin radial and spiral bamboo rods, it represented a radical and highly innovative new approach. The cabinet which housed the actual loudspeaker featured beautiful and ornate fretwork, designed by former draughtsman Ralph and destined to become the signature look of a blossoming Celestion product line.

The Celestion was favourably reviewed in the predominant journal, *Popular Wireless and Wireless Review*, in its 'Apparatus Tested' section of 25 April 1925, as 'a high-class instrument capable of high-class performances'. The journal prefaced this statement by saying it had been 'one of the most interesting instruments we have had the pleasure of testing', describing it as 'artistically as much in advance of the horn type loudspeaker, as the hornless gramophone is in relation to the old-fashioned phonograph'.

In appraising its sound quality, the reviewer was impressed by its transparency. 'In operation results are extraordinarily fine, and "strings" come through with all their original timbre, while other instruments and speech are most faithfully reproduced.'

'The Celestion is very sensitive and it does not require anything but ordinary LS [loudspeaker] signals to operate it successfully', the review continued. 'Its sound-projecting qualities are excellent, and it can be pushed to the limits of LF [low frequency] plus power without distortion arising.' A similarly positive review appeared in *The Wireless Engineer*.

It clearly captured the imagination of the public, for, in the same month—April 1925—*The Broadcaster & Wireless Retailer* carried an announcement from one of Celestion's Belfast dealers which declared, 'The Celestion loudspeaker is the latest agency acquired by Messrs. Dobbin, of North Street, Belfast. This instrument can be had in either oak, mahogany, or walnut, at prices ranging from £6.10s to £6.15s. It is guaranteed for 12 months. Many have been sold already.'

Radio had truly become the medium of the age, and as a result of the 'French Revolution', Celestion was set fair for success

A second patent, filed jointly by French and Mackintosh the following year, described an improvement to this design. It demonstrated that clamping the loudspeaker cone rigidly around its outer edge had the effect of creating a larger vibrating area (hence improving sound reproduction), finessing the earlier *free-vibrating edge* paradigm. It was finally issued on 14 January 1926, as British Patent No. 245,704.

That second innovation—the fixed edge—was a genuine improvement to Eric's first design. It was crucial in enabling the loudspeaker to function more effectively, allowing the company to offer genuine performance improvements in models that were to come soon afterwards.

By 1926 the Celestion Radio Co. had been created as a separate entity to manufacture Celestion loudspeakers and the popular Celestion Woodroffe gramophone pickups which had been quickly added to the range, and by 1927 it had registered adjoining premises at 31 to 37 High Street. In 1928 the Celestion Radio Co. was also listed as an occupant of 20 Union Street, Kingston-upon-Thames, where it manufactured 'Wireless Accessories'.

It appears Celestion was also active on the early audio trade exhibition scene: presenting at the Wireless Show at London's Olympia in 1926, their entry in the show programme was as follows:

> The 'Celestion' was one of the first cabinet type loudspeakers in which the sound is emitted from a large conical diaphragm. The driving armature, which is operated by a compound magnet system, is attached to the centre of the diaphragm. A very thin parchment-like material is used for constructing the cone which is stiffened by a spiral. The case represents a high-class cabinet work and is obtainable either in oak, walnut or mahogany. Although it is difficult here to make comment on the performance of a loudspeaker, its merit is evidenced by the high reputation it has achieved in the comparatively short time that it has been on the market.

Meanwhile, at the Wireless Exhibition that same year, they presented a speaker 'with a diaphragm 28in across' (something so large could only have functioned thanks to the rigid outer-edge clamping innovation).

Left: An excellent example of the mahogany C14.

Right: Advertising pamphlet featuring the C14 radiophone.

Such rapid expansion supports the claims that the original 'Celestion' had been a popular seller, ensuring that the business got off to a flying start. It also gives credence to the (uncorroborated) story that Ralph and Eric were given a bonus of £5,000 each, a considerable sum of money in those days!

Celestion speakers quickly became an essential component used with the new valve receivers that would appear as a result of the BBC stepping up its transmitter power. With transmitting stations also starting to appear in Europe, it became possible to locate and pick up programmes via the new, more selective receivers. Radio had truly become the medium of the age, and as a result of the 'French Revolution', Celestion was set fair for success.

2 The Very Soul of Music

At the end of 1928, flying under the enduring slogan 'The Very Soul of Music', the Celestion Radio Co. was folded into the newly incorporated Celestion Ltd. The original Celestion radiophone had already been followed by a range of extension speakers, with the A1, A2, A3, A4 and A5 quickly coming into production. Boasting 'British Manufacture Throughout' and guaranteed for 12 months, the advertising literature of the time highlighted many technical aspects of the speakers that contributed to making them 'The World's Super Radiophone'.

Technology and design were already important selling points for the Celestion brand, as the A Series speakers were documented as having a 'Non-Resonant System with No Directional Effect' as well as boasting 'an elegant appearance and accurate workmanship'. All five products featured the same bamboo-reinforced cone loudspeaker seen in the design that Cyril and Eric had patented, housed in ornately carved cabinetry available in oak, mahogany, walnut or ebony.

As the audience's needs had grown greater and the radio technology improved, so the company had quickly diversified. The Celestion Self-Contained Balanced Signal Radio Receivers which followed allowed 'the very personalities of the artistes to enter your home', bringing a radio together with the speaker cabinet in a stylish mahogany case.

By 1928 it wasn't just finished cabinets that were being sold, as the company began supplying other manufacturers as well, including selling cone loudspeakers to the renowned Pye electronics company. Cyril French signed a contract for 10 years as managing director and, with trade burgeoning, a move to large-scale manufacturing premises was inevitable. Thus, in 1929, Celestion Ltd moved from Hampton Wick further along the river to a much larger factory at 145 London Road, Kingston-upon-Thames.

Territorial expansion saw Cyril and Ralph French both journey to Paris, where they came to an agreement with S.A. des Établissements Constable, a company formed by Major Henry Constable Roberts, and by December 1928 they had set up Établissements Constable-Célestion, with

Opposite: The Celestion Radio Gramophone.

Roberts consequently joining the board of Celestion Ltd. Products were exported from Croydon Airport, the UK's only international airport at that time. It's likely to have been one of the earliest examples of commercial export by air.

For sales purposes, showrooms were maintained in Villiers Street, Charing Cross, and nearby Victoria Street in Central London, as well as Warren Street just a few blocks west of the City Hall in New York's Manhattan. And in addition to Constable in France, they had also appointed Krischker & Nehoda in Vienna, Austria, and further afield, Veal & Co in Melbourne, Australia, as distribution agents.

December 1928 was also the month when Celestion was formally registered as a public company, enabling shares to be bought and sold on the London Stock Exchange. Sir Louis Stanley Johnson was installed as Chairman of the Board, and Cyril French was given a 10-year contract as managing director. This was in no small part due to the profitability of the company, which had shot up almost five-fold, from £11,063 (year ending February 1927) to £51,744 (year ending February 1928).

Following the success of the A Series, Celestion then produced the C range of elegant mahogany radiophones—C10, C12, C14 and C24—which were brought to market with prices ranging from £5.10s to £25. The C10 and C12, in particular, were highly praised. The former was claimed to be 'the finest loudspeaker procurable at the price' and the *Popular Wireless* issue of 31 March 1928 described the C12, which incorporated a cobalt magnet, as a 'long way ahead of its class'. The review went on to say:

> To obtain a better speaker than this model C12 one would have to pay very much more money. As a matter of fact, a move up into the moving-coil class would, in our opinion, be necessary. The C12 is a cabinet cone and its construction and finish are so good that one stamps it 'deluxe' at a glance. A point in its favour is that it is not particularly large, being only 6in. in depth, and 14in. x 14in.
>
> It has a splendid projection, every vibration coming cleanly away from the cone, and we know of but one or two other cones which have less coloration. Speech is decidedly of moving-coil quality. There is no appreciable resonance and both bass and high notes are faithfully dealt with. The speaker is rather more sensitive than the majority, but it is capable of handling heavy inputs. We found it perfectly satisfactory on each of the several sets with which it was tested, ranging from two valves to a multi-valver of the super kind.

By 1929 Celestion were boasting of a considerable global outreach. An edition of *Broadcaster & Wireless Retailer* bore an advertisement for the C12 and C14 that demonstrated that export ambitions had been fulfilled, under the headline: 'In every part of the Empire.' It read, 'As far South as New Zealand, as far North as the Hudson Bay, East to Hong Kong and West to

Left: C12 advertising, *c.* 1928

Right: A fine example of the mahogany C12 radiophone.

Vancouver, British Celestion enjoys the reputation of being the standard of comparison amongst loudspeakers. Celestion sales have soared steadily but surely, even during the recognised off seasons. Sheer merit and perfect craftsmanship aided by the Celestion patented reinforced diaphragm have earned this enviable distinction'.

1929 also coincided with the arrival in the company of both W. 'Billy' Page and S.J. 'Jim' Tyrrell, who would both eventually join the board 17 years later and, in the early 1950s, become joint managing directors.

However, the Great Depression was soon to ripple across the globe, instigated by New York's Wall Street Crash. This cataclysmic event would herald years of mass unemployment, economic hardship and hunger marches, touching everybody's life. Between 1929 and 1932 British industry suffered terribly. Export trade fell by half and the output of heavy industry fell by a third. Demand for shipbuilding plummeted by 90%, cascading through hundreds of supply chain industries, so that by summer 1932 unemployment had reached 3.5 million, a significant portion of the working population.

By February 1930 there were signs that Celestion might have been preparing for a downturn. A formal letter from Cyril French to brother Ralph, who at the time was Head of the Buying

Ralph French working in the showroom office, Victoria St., London.

& Records Dept, confirmed a reduction in his salary in exchange for shares following a company reorganisation. Cyril added the promise, 'You may rest assured if at the end of 12 months there is a revival in trade, we will review the salaries with reference to an increase for those who have assisted me'. Sometime between 1930 and 1932 Ralph appears to have left the company, although the exact circumstances of his final departure remain unclear.

Perhaps as a further indication of the difficulties to come, records show that as of 1 October 1931, the subsidiary in France was put into compulsory liquidation. The operation was considerably downsized and transferred to an address (possibly a sales showroom) in Neuilly-sur-Seine, just outside Paris, in June 1932. A judgement on 30 April 1934 declared the Constable-Célestion operation bankrupt for lack of assets.

In the time immediately after the change of premises, however, the Kingston factory had continued to build up a head of steam and there appeared to be no immediate brakes on progress. While the company had been growing throughout the second half of the 1920s, Cyril and Eric both kept the Patent Office busy, filing numerous separate patent applications between them, with descriptions relating to improvements in 'acoustic apparatus' and 'sound reproducing instruments'. Ralph and youngest brother Edgar also appeared to be active in company research and development, having filed patent applications under their own names.

One can imagine that the early years of Celestion were fired by an inventor's zeal, as the company strove to improve and build on its early successes with even better and more innovative products, even as recessionary headwinds were building.

In 1931 the C range of loudspeakers was supplemented by the new models D10, D12 and D50. Technological improvements meant Celestion needed to stay up to date. The first electrical disc-playing machines had appeared in the late 1920s, and these electric 'phonographs' became more widespread and were later combined with a radio receiver. By the early 30s receivers had become more sophisticated, and smaller loudspeakers were now being built into the receiver cabinet itself, rendering a separate unit unnecessary.

Celestion continued to develop quickly in both the mechanical and electrical spheres of the home entertainment industry, for which they produced luxurious gramophone products. The units became so popular that Celestion issued separate catalogues for Electric Gramophones, Reproducers and, rather oddly, Band Repeater Equipment for ships. Produced with a fine vessel on the cover, steaming with porthole lights gleaming over a tranquil sea, one interesting item illustrated is the Type S/LSP loudspeaker, which must surely be one of the first dual-purpose public address loudspeakers. It seems difficult to appreciate the significance of the caption nowadays, which stated that the loudspeaker unit could be bracketed to a bulkhead, above or below deck ... 'remote from the instrument'.

One booklet also proudly illustrated three grand models of electrical and radio gramophones in craftsman-built footed cabinets of oak, mahogany and walnut, now priced at between 75 guineas and 125 guineas [a guinea being equivalent to £1.1s]. The electrical gramophone incorporated a positively silent motor; apparently the world's finest electrical pickup: the famous Celestrola moving-coil loudspeaker; a completely screened amplifier; and continuous control of volume from maximum to zero. All were contained in a richly designed and finished oak or mahogany cabinet. The sales literature promised that 'its realism is literally breath-taking and every delicate shading is recreated. Distinctive features are a clean and solid bass, very high frequencies without resonance and extreme sensitivity.' Its companion, the Celestion Radio Gramophone, incorporated the same features, but added a completely screened radio unit with the latest high-frequency (HF) valve for long or short wave, the radio being made available to the listener by means of a single switch.

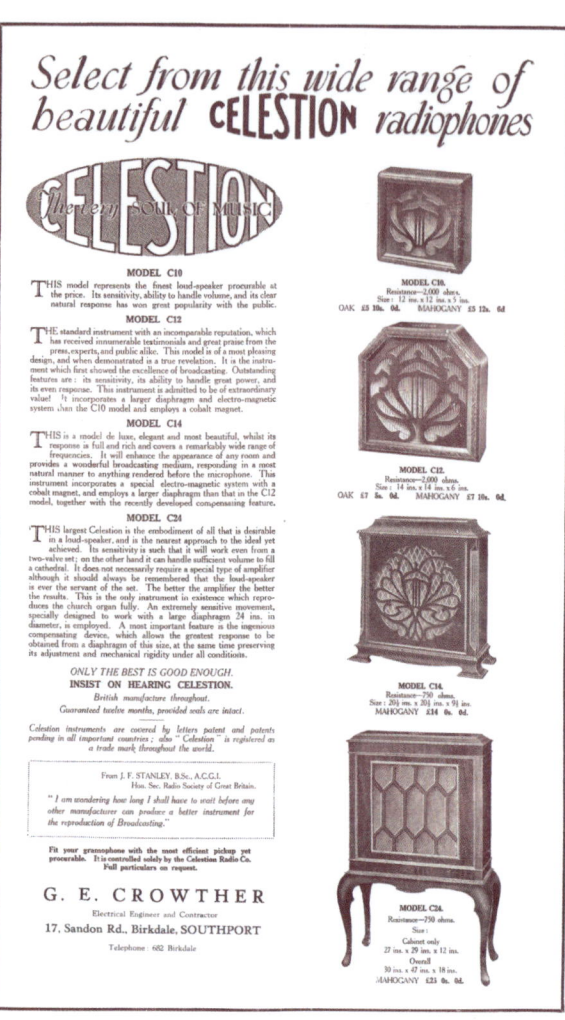

Left: A magnificent example of the mahogany C24, complete with 24" cone speaker.

Right: C Series advertising pamphlet.

In 1931 the company also introduced a 'Recording Gramophone', supplied complete with blank disc and cutting needles! Operated using the 'Kingston method' of recording (invented by one Arthur Kingston), according to *Wireless World* it was 'provided with a good motor that allows maximum weighting on the cutting needle when recording so that a deep groove is obtained'.

Meanwhile, with the demand for mains receiving sets—which overcame the disadvantages of batteries and accumulators—requiring mass production, styles became less ornate, the fretted front being replaced by a cloth or metal grille. A clear case of function taking precedence over form.

By the early 1930s the company was ready to present its first venture into a PM moving-coil speaker, the PPM. Until this time permanent magnets had been bulky and expensive, so most loudspeakers were energised using externally powered electromagnets. In October 1932

The Broadcaster & Wireless Retailer produced a complimentary article under the headline, 'Improved Celestion PPM Standard Speaker', so by this time the PPMs had been successful enough and been around long enough to warrant updating. The article read:

'Several improvements have been made in the construction of the PPM Standard permanent-magnet moving-coil speaker manufactured by Celestion, Ltd., of London Road, Kingston-on-Thames', the article read. It continued:

> An improved universal transformer will in future be fitted, allowing the speaker to be accurately matched to both power and pentode valves. Also, an improved "Hyflex" moulded diaphragm will be used in place of the present type. An outside suspension will replace the internal metal suspension used at present, and a special speech coil with a moulded former will give added reliability. The improved model has been renamed the PPM19, but the price will remain at £47s.6d. The PPM19 was released today (Oct 1), and all orders in hand will be satisfied by this improved model.
>
> The result of seven years specialisation the famous PPM speakers comprises not one, but a complete range, each speaker a masterpiece of design and construction. Celestion's reputation is already well established, but their latest achievements, which are far in advance of anything they have yet produced, greatly enhance their popularity. Remarkable tonal quality, volume and sensitivity are the most outstanding features of the PPM range. On performance alone these speakers would sell, but backed as they are by an extensive advertising campaign, success is ensured.

'Stock and sell these wonderful speakers and make certain of having a bumper season', was its final word of advice.

At the same time, the Reetone, the first-ever dual speaker system, was built, offering a generous power rating and leather cone suspension. *Wireless Trader* remarked that 'it handled 2 watts easily ... its tone was very good indeed ... bass generally was firm and clear and not booming, middle register was even and free from peaks, while upper register was well in evidence and clear and crisp'.

The Reetone principle involved two speakers being built into a fascia plate, each with its own input transformer. The Matched System featured two speakers that were of equal size and in most ways the same, but had mechanical resonances that were 'staggered'. According to the advertising, this 'eliminates the tendency to "boom" ... and generally augments the output below 150 cycles'. The Dual System, on the other hand, featured cone speakers of two different sizes, coupled so that the high frequencies were accepted by the smaller treble unit and the lows by the larger bass unit. Very much a precursor to a home HiFi speaker.

The two-speaker system was among many ingenious ideas incorporated into new designs intended to raise the quality of sound reproduction. *Practical Wireless*, in its issue dated 28 January 1933, described the Dual System as having 'remarkable lifelike tonal quality and sensitivity; for with no other speaker is the bass and treble reproduced so naturally and free from distortion ... with such vivid realism'.

Elsewhere, it was announced that a new vaudeville-type act styled 'The National Broadcasting Contest' would shortly commence a tour of British cinemas. In this show members of the audience would be invited to give a performance of their own with a view to discovering fresh talent for broadcasting. The 'turns' would be relayed to the theatre audience by means of a three-stage amplifier and speaker. The speaker was the new Celestion *Auditorium* model, while Celestion Ltd had constructed the amplifier to the organisers' design.

By this time, the Kingston-upon-Thames factory housed a sizeable workforce, manufacturing the full range of Celestion products. There was a dedicated assembly line for mains-driven electromagnet-type moving-coil speakers and another for assembling cabinet models. An extensive coil-winding department was staffed almost exclusively by women, who deftly operated the automatic winding machines.

A transformer assembly line was involved with the manufacture of multi-ratio transformers, enabling several current ratios from a single transformer. These were for use in conjunction with the various speaker models available. All Celestion speakers were designed onsite in a 'most efficient and up-to-date laboratory' which was 'staffed by highly skilled engineers of many years' experience', reported the *New Era Illustrated* in February 1933.

By 1935 the company finally fell victim to the fallout from the Great Depression. With an announcement of reorganisation, Celestion's first 'boom' era was finally, officially, ended, as Cyril French resigned from the board of directors, with one R.B. Page being installed as the new managing director in his place. No record exists of the fate of co-founder Eric Mackintosh, but he appears to have left the company in the early 1930s, around the same time as Ralph French.

Cyril returned to the original, still-functioning Plating Company building at 29 High Street, Hampton Wick, to operate as 'sole wholesale and retail distributor of Celestion products in Great Britain', which he was to do successfully for many more years.

Clockwise from top left: The Kingston-on-Thames factory, street view; the factory, aerial view; assembling transformers; assembling cabinets; the machine shop; voice coil winding; the engineering laboratory.

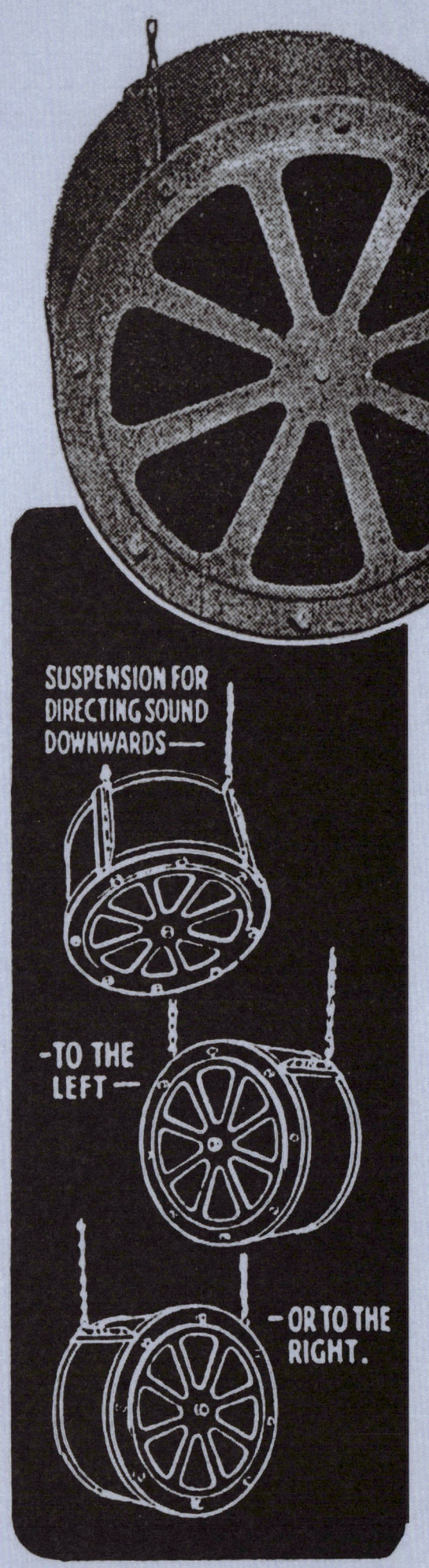

ADDRESS YOUR STAFF THROUGH A CELESTION LOUDSPEAKER

with a volume sufficient to overcome local noise

The same speaker will carry music to your staff — true music, not just a volume of sound

The Celestion method of suspension enables the sound to be directed wherever you want it

3 Recession, War and Rola

Through much of the 1930s, British Rola, an offshoot of the Rola Company of Cleveland, Ohio, had been operating almost in lockstep with Celestion. While Celestion was based at its plant in Kingston-upon-Thames, in North-West London British Rola had been making similar products, located first in Kilburn High Road and later at a nearby factory in Minerva Road, Park Royal.

By 1933 half the households in Britain owned a radio, and Rola and Celestion competed for the home and export markets, their products influenced by the same changes in the wireless receiver market. As the receiver became more sophisticated and smaller, the loudspeaker began to be more often housed within the receiver cabinet itself, rendering the separate speaker unit unnecessary. Products were often developed to a client manufacturer's specification—and this was the fertile market being chased by both companies.

Rola had quickly shown itself to be a force to be reckoned with in the world of loudspeakers, producing ranges of radiograms and extension cabinets as well as chassis speakers without enclosures, and in 1936 had launched the G12 PM loudspeaker. Very well received at the time, this model was to take on a whole lot more significance later.

Despite the rocky economic climate, the work of loudspeaker development and production continued unabated at Celestion in Kingston. A Celestion catalogue from around 1936 (with Cyril French listed as sole distributor) laid out chassis models for sale to other cabinet manufacturers. This ranged from the huge P84, 18in speaker with an enormous peak power capacity of 40W and priced at £25.4s, to the tiny 2½in P2V0 offering a modest ¼W at £1.7s.

Also on show were the now more streamlined walnut, oak and mahogany bookshelf cabinets, 'suitable for all receivers', such as the Junior Auditorium model fitted with volume control and available with transformer (part number CT113, priced at £6.15s) or without (part number CT114, priced at £6.9s). So, the loudspeaker had not vanished entirely into the integral receivers —indeed, many purchasers still preferred their speaker to be separate from the receiver.

Opposite: Wartime advertising.

The new Celestion Amphenol valve holders are the strongest holders in the world today, and yet are compact in size, modern and attractive in appearance and lower in price

Records show that by 1937 Stephen Philip de László had taken over the reins as Celestion's managing director. Second son of the Hungarian/English painter Philip Alexius de László, the socialite and entrepreneur had a great interest in the burgeoning wireless technology. He had previously gained experience in electronics while working at Raytheon in Boston, Massachusetts, and after returning to the UK set up Hivac, a manufacturer of vacuum tubes, with his younger brother, Patrick (Hivac was later sold to English Electric).

Stephen de László's tenure at Celestion was brief. As a result of his untimely death in a car accident in December 1938, he was succeeded as managing director by brother Patrick, who was to oversee company operations right the way through the Second World War.

By 1938 British Rola Ltd was registered as electrical and general engineers, toolmakers and stampers—the company's motif being a picture of a bird on a branch with the words 'The speaker you know by ear'. When the Second World War broke out the following year, the company became supplier to the British Air Ministry, opening up a dispersal factory at Bideford in Devon, in the former garage of Messrs Elliot and Sons.

Production for the war effort grew apace; the official list of suppliers shows British Rola making 'aircraft vacuum pumps, fuel pumps, de-icer pumps; also, precision production items to Air Ministry or contractors' specifications, tool making, pressings and stampings'. They were involved in the manufacture of the RAF B3 Vacuum Pump, several thousand of which were produced. These were followed into production by the Rotol Airscrew Feathering Pump, four of which were fitted to British multi-engined bomber aircraft, as well as the Integral Hydraulic Pump BH Mark IV. All significantly helped the British and Allied Forces to meet their need for reliable battle equipment.

Celestion was not on the official suppliers list. Prior to the Second World War, they had begun making radio accessories, adding such items as 'MIP' (moulded in plate) radio valve holders to its impressive product range. These were advertised thus: 'The new Celestion Amphenol valve holders are the strongest holders in the world today, and yet are compact in size, modern and attractive in appearance and lower in price.' Alongside these were the Celestion Amphenol Microphone Connectors, which were marketed in both English and US types and, with applications beyond that of domestic audio, continued to be made for the duration of the war.

WALNUT

OAK

MAHOGANY

Left: A fine example of the Junior 8, streamlined cabinet.

Right (top to bottom): A Junior Auditorium in walnut. A Junior 8 in oak and a Standard 8 in mahogany.

At the same time, much of the remaining production at the Celestion factory changed to munitions work, which included, for example, the soldering of bomb holders for aircraft. Both Celestion and British Rola were reportedly restricted to producing only one type of loudspeaker, the utility 'W' type, from their respective manufacturing bases. Around 1942, an assembly shop was built by Celestion on the field of the Tiffin School in Kingston, directly across London Road from the factory.

According to school archives, there were also air raid shelters below the assembly shop that had been built at the beginning of the war by Celestion, and it is documented that some of the shelters were shared by boys from the school and Celestion workers as the sirens wailed above. By 1948 the Celestion building on the school field had been handed over to the school (as had been agreed at the outset) and became the school gym.

In 1942, Rola's US parent company, which had been founded in 1924 by Bernard A. Enghorn, transferred to British Rola the right to manufacture and sell throughout the British Empire (excluding Canada and Australia) and the Continent of Europe such products as were used in

RECESSION, WAR AND ROLA

Left: late 1930s advertising shows the 18" P84 next to the 2½" P2VO.

Right: Portrait of Patrick De László painted by his father, Philip.

the aircraft, engineering, electrical, motor and radio industries. At this time the Rola managing director was R.W. Cotton, and on 4 February 1944, he joined the board of Philco Radio. It was hoped that as both companies had US affiliates, and both made products for the radio and allied industries, they would work in close co-operation, though this partnership never came to fruition, as Cotton left Rola soon afterwards to manage the newly formed Philco Overseas instead.

Patrick de László, meanwhile, had been drafted by the Royal Air Force, at the rank of Group Captain, with a mission to build a radar chain around Britain, which was crucial in directing Fighter Command. This culminated in the successful concentration of airborne resources in the Battle of Britain (the successful defence of Britain against air raids conducted by the German air force from July to September 1940, after the fall of France).

According to one source, Celestion Ltd was the company chosen by the Government Research Establishment (GRE) to bring the laboratory prototype of the Proximity Fuse that the GRE had developed into full production. Essentially a miniature radar transmitter and receiver operated

Cutaway of a proximity fuse.

by a chemical battery, the fuse became an integral part of an anti-aircraft shell, enabling each shell to detonate when it came within lethal distance of its intended targets, which were the Nazi V1 rockets that had been relentlessly attacking Britain during 1943.

De László was released from active work in the Air Ministry to supervise this complex and successful project, eventually setting up the company Micro Precision Products on the Celestion site at London Road to concentrate solely on the task. The final designs were then sent to the United States for mass production, and the fuses played a vital part in countering the worst of the V1 bomb attacks.

As the fog of war finally lifted in 1945, Celestion and Rola both emerged intact and attempted to resume more normal business activities. However, Britain had been indelibly changed, and continued survival in a new era of post-war austerity meant there would be drastic measures to come.

The Rola Model G12

It was 1936 when Rola introduced its G12-format 12-inch diameter speaker models. A version with a direct-current (DC)-energised electromagnet was supplied, either complete with a mounting stand, handle and base or 'stripped' (i.e. without the accessories). A PM version featuring a 'Nickel Aluminium Cobalt magnet' (alnico in modern parlance) was produced, and both DC and PM variants were available with or without a transformer. As the literature of the time had it:

> When Rola technicians set out to design the finest speaker available to the general public, they produced the Rola G.12—a full 12" diameter High Fidelity unit with performance that outclasses that of any reproducer outside a laboratory. As with all Rola models, every component of the G.12 is made by Rola ... the speaker is built, not merely assembled at the Rola factories.
>
> In the G.12 P.M. a Nickel Aluminium Cobalt magnet of exclusive Rola design is employed, whilst the Rola cone used for this model is more sensitive and gives greater overall response than any other. The G.12 gives a very uniform response from 50 to 12,000 cycles—emphatically it is the speaker for the connoisseur. Hear one yourself and let your own ear substantiate our claims.

It was greeted with excellent reviews, such as this one from the March 1936 issue of *The Gramophone*:

> This permanent magnet version of the G.12 loudspeaker is, without doubt the most efficient of its type that has so far passed through our hands. In the matter of quality of reproduction, too, we do not remember having had a permanent magnet moving-coil speaker that has given us so much satisfaction.
>
> The magnet, a circular one some $4\frac{1}{2}$-ins. in diameter is particularly powerful. This feature is, no doubt, responsible to a large degree for the high efficiency ratio of output to input of the speaker, as well as having an important effect on the quality and range of frequencies reproduced, especially bass frequencies and transients. These are rendered remarkably well. The [DC] energised G.12 can still claim some superiority in the matter of depth of tone, but the advantage is only comparatively small.
>
> On the other hand, we are not sure that this PM. G.12 does not give a livelier result; the attack is virile and the definition is wonderfully clean. These characteristics prevail, too, at relatively small inputs. This, in our opinion, is one of the most valuable features of the speaker. There are no peaks of sufficient magnitude to discolour the reproduction; not even large inputs seemed to disturb the speaker's equanimity.

It would be another 25 years before the G12 would be catapulted to stardom, giving voice to rock 'n' roll and the amplifiers that powered it. But even in 1936, when that particular application could barely have been conceived of, the elements of the 'modern' G12 were already in place. Note the shape of the chassis, crisply stamped, the tall alnico magnet and use of a bell cover. Over a long period of time the G12 has proven itself to be a legend of the audio industry. This is its origin story.

When Rola technicians set out to design the finest speaker available to the general public, they produced the Rola G12

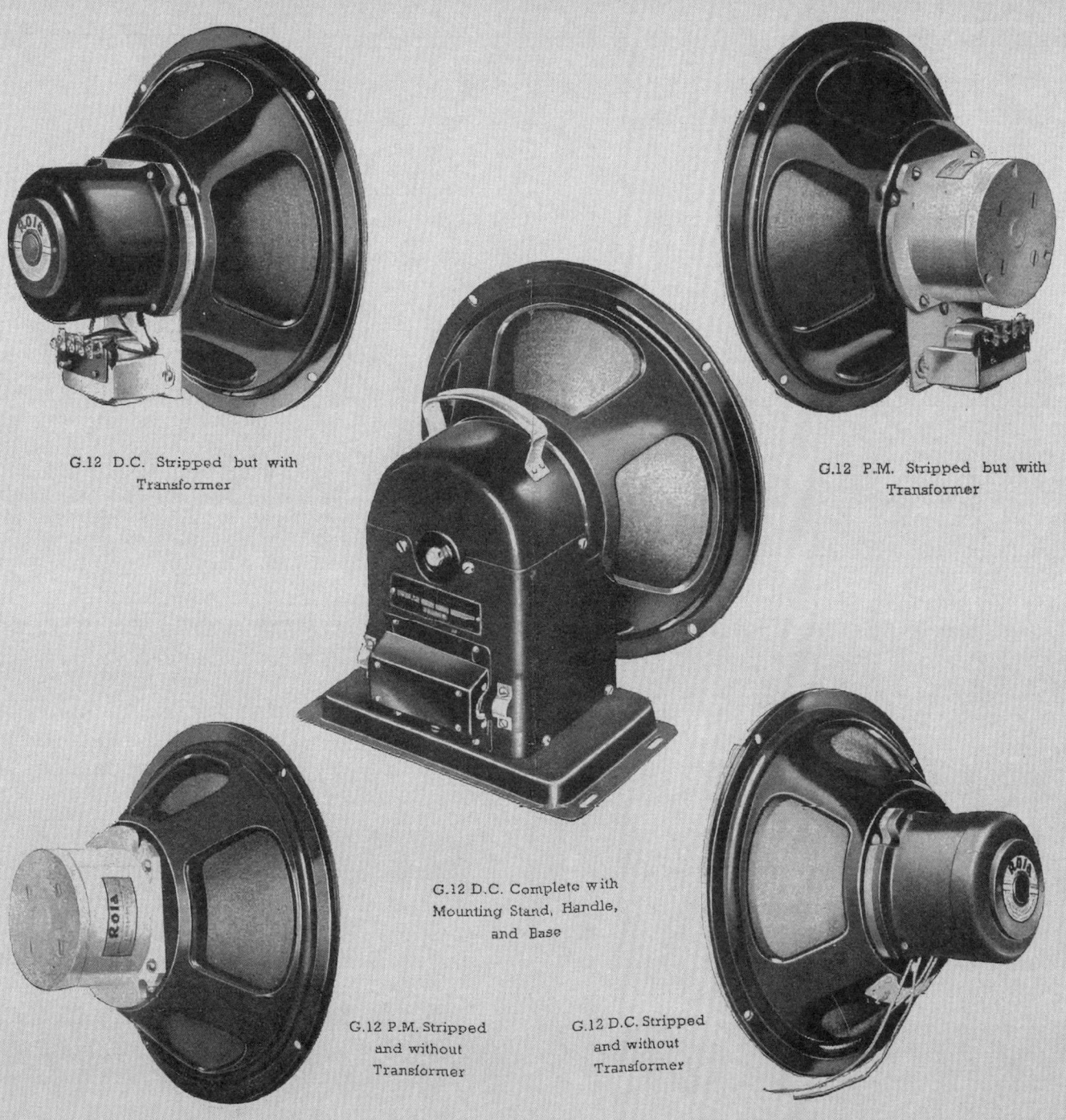

G.12 D.C. Stripped but with Transformer

G.12 P.M. Stripped but with Transformer

G.12 D.C. Complete with Mounting Stand, Handle, and Base

G.12 P.M. Stripped and without Transformer

G.12 D.C. Stripped and without Transformer

The G12 range in 1936.

AH! BUT THAT ONE WAS A ROLA

Rola SPEAKERS
SELL THEMSELVES ON THEIR REPUTATION

...and the Receivers in which they have been installed!

4 Rola Celestion

On 31 July 1946, in a bold move for rapid growth just as post-war austerity was beginning to bite, British Rola acquired Celestion, eliminating a significant competitor and increasing exports in one fell swoop. At the same time, they agreed to purchase several subsidiary firms: Pressmach, based in Woolacombe, Devon, another small pressing firm, Belark and a furniture company. The *Investors' Chronicle* for 1946 reported that following the takeover, British Rola would be responsible for the production of considerably more than half of the loudspeaker trade in the UK, and the two companies together practically covered the country's entire export loudspeaker business.

With the coming of peacetime, the British Rola dispersal factory in Bideford had closed down after all government contracts had been terminated. Consequently, personnel and speaker assembly were transferred, joining British Rola's London operation at a newly acquired premises in Thames Ditton, a mere three and a half miles south-west of Celestion's Kingston home.

Because of the lack of building materials, development progress at the Thames Ditton facility was slow, so the British Rola factory in Devizes, Wiltshire, remained open to ease pressure. As the company prepared for the takeover of Celestion, they went to the market to raise and issue 825,000 shares valued at two shillings, increasing capital to £175,000.

The factory on Summer Road, Thames Ditton, also known as Ferry Works, certainly had some provenance. Originally constructed by the engineers Willans & Robinson in 1879, much was rebuilt after a disastrous fire in November 1888. The reconstructed building utilised a 'saw-tooth' north light roof, the earliest known example of the application of this technique to the construction of a machine shop. In 1911, after the firm had moved to Rugby and been absorbed by the English Electric Company, the premises were taken over by Auto Carriers Ltd, makers of the well-known AC cars.

Immediately prior to British Rola moving in, the Ferry Works site had been occupied by Astor Engines, a manufacturer of steam engines who had used their own generators to power the plant. It is believed that this site was the first in the country to be illuminated entirely by electricity.

Opposite: Rola advertising, 1946.

Out in the commercial world, in the immediate aftermath of the Second World War, the improved version of the black and white television had resulted in this medium becoming more commonplace in homes. And while the radio speaker sector remained strong, this new market was destined to become a principal focus of loudspeaker manufacturers.

The takeover was formalised in 1947, and production continued in earnest under both the Rola and Celestion names. British Rola was listed as an exhibitor at the British Industries Fair, which was held at Olympia and Earls Court, London, and organised by the Export Promotion Department of the Board of Trade. On the ground floor in Olympia, they showed a range of six speakers for all types of radio, both 'energised (electro-magnet) or permanent magnet type', with sizes ranging from 5in to 10in.

At the same time, the fair carried a Celestion advertisement for the 'Queen of Loudspeakers, Specialists in Loudspeakers of all sizes (with and without Transformers) for Radio Set Manufacturers, Relay Services, Inter-office Communications, Extension Speakers for the Home and other purposes'. An advertisement the same year in Practical Wireless showed Celestion-branded chassis being produced, with diameters ranging from 2½in to 18in.

In the summer of 1948, Celestion vacated the Kingston-upon-Thames factory, and production machinery and personnel moved to Thames Ditton. The two company names became braided together as Rola Celestion Ltd, and Celestion was adopted and registered as the trademark for the company's product.

Engineer Bill Strong—who joined Rola in 1946 after demobilising from his Army deployment in the Far East—recalled, in an archive recording, the structure of the company after Celestion moved in. 'Several of the Celestion management became Rola Celestion management and the MD was Jim Tyrell, a very nice man. Billy Page was sales director and Arthur Young came in as technical director. We had Bill Wren, who wasn't a Celestion man and he was the financial and office manager. Then there was Fred Brightwell, who came in as works manager and Harry Coleman continued in his position as tool room and engineering.'

Strong confirmed that prior to the Second World War British Rola had still been based in Park Royal. 'For the whole of the time we were at Ditton from the end of the war a daily van used to come from Park Royal bringing people who had been pre-war employees. In fact, when I started at British Rola [at the end of 1946] it seemed to be a fairly well-established operation [at Thames Ditton]. We needed to expand but the problem was there was nowhere to expand into. Ferry Works was completely surrounded. Houses on one side, a pub on the other side and the other two sides were the river and the road. At the time we only had half the premises, the other half was occupied by Maurice Jelinek who made hairbrushes.'

Top: The Rola Celestion factory in Thames Ditton.

Right: The announcement of a 'merger' between Rola and Celestion.

> they have been clearing up following building work on the factory, and making wireless transformers with the help of a concrete mixer driving an air compressor

The extremely hard winter and fuel crisis of 1947 caused havoc to production, and two interesting extracts from the *Surrey Comet* newspaper for 15 February illustrated this problem: 'In an effort to get production started again, Celestion ... have bought petrol driven generators, but yesterday they were still trying to obtain permission from the Ministry to use them and to get the petrol to run them. If they are successful, they will be able to bring some of their 240 workers, most of whom are women, back on the job, but at the moment production is at a complete standstill.'

And then ... 'Over 300 employees at British Rola's Ferry Works, Thames Ditton, will continue for a further week to receive guaranteed wages as a minimum. A previous decision to close the factory if power were not obtainable on Monday has been cancelled because, according to Works Manager Jack Jones, the staff have responded magnificently to an "all-hands-on-deck" appeal by the directors. They have been clearing up following building work on the factory, and making wireless transformers with the help of a concrete mixer driving an air compressor. Ninety per cent of loudspeaker output is for export.'

However, the new company hit problems before it really had a chance to get started when the creditors came calling. The aftermath of the post-war fuel crisis and the slow fit-out of Ferry Works, as well as lower-than-expected loudspeaker sales and unexpectedly low profits from the allied furniture company, meant that there was not enough cash in the coffers. By June 1948, with £54,000 required to clear the debt, the receivers were called in.

Bill Strong recalled, 'eventually in 1948 a pretty serious financial problem arose and large numbers of staff and employees were laid off and I was one of them. We were called in on the Friday and told there was a very serious problem and they were sorry. They gave us a week's money *in lieu* of notice and that was it. But in my case, about a fortnight later or less a chappie came round from the company and said would I be ready to start back again? And I started back a fortnight after I left.'

On Strong's return, he immediately noticed a number of changes. 'A lot had clearly been happening,' he said. 'To start with, works managers Jack Jones and Mr Loach had disappeared and in fact they went to Australia and set up a speaker operation there. Mr Poole had left the company and taken over the Devizes factory and bought all the transformer manufacturing equipment and he was setting up Hinchley Engineering ... to produce [small electrical] transformers.'

To find additional money, there was a 'realisation of assets' that included the sale of property and equipment, including Celestion's London Road factory and much of its contents, as well as the Rola factory in Devizes. This was the catalyst that finalised the merger of the two companies under one roof.

Despite best efforts to raise the capital required, it was still not enough to reimburse both preference and ordinary shareholders, and the company remained indebted. By July 1949, with no turnaround in sight, Rola Celestion had little choice but to enter voluntary liquidation.

Insolvency specialist and liquidator Cork Gully was brought in to handle the sale. Bidders were said to have included Plessey, Murphy Radio, Pye and Echo, but it was Truvox, the company chaired by the multi-talented, if somewhat mercurial, Daniel Dan Prenn, that was to prove successful. The bid was said to be for £135,000 'with conditions', but after finding 'irregularities', £30,000 of that capital amount was supposedly returned to Prenn.

Rumour had it that Prenn, whose head office was situated at Mount Street in London's affluent Mayfair neighbourhood, had a friend who was an official receiver that had recommended companies worth buying. With so many diverse business interests already, however, would he be the man to lead Rola Celestion towards a better future after merging it with his Truvox Engineering business?

Truvox, based in the Wembley area of North London and with an additional manufacturing operation in Wales, was well known for its public address speakers and systems that included horns and loudspeakers for cinemas as well as acoustic devices such as re-entrant horns that the company had developed for wartime use.

In November 1949 Rola Celestion's survival was finally guaranteed when the deal was completed and the company was bought out of receivership. Truvox's public address loudspeaker systems would soon be folded into the diverse and growing product range of Rola and Celestion, and for the next two decades it was under the Truvox umbrella that Celestion would flourish.

1950–1970

2
Transformation and Evolution

The Beatles perform at the BBC.

5 Truvox and the Empire of Daniel Prenn

In buying Celestion and Rola, Daniel Prenn's rationale was to take out two major competitors in a single move, with Tannoy and Goodmans being the other big business rivals at the time. The two companies were thus folded into the Truvox family in November 1949, the decision finally becoming official on 20 February 1951, with the resultant company retaining the name Rola Celestion Ltd.

The two decades that Rola Celestion spent as part of Truvox were diverse and successful, in many ways befitting the entrepreneurial flair of the company's colourful owner and driving force. But who exactly was Prenn, and what was the story behind Truvox?

Born in 1904, Daniel Prenn had often been described as Russian; however, other records indicated he was born in Wilno (or Vilna) a city that had been claimed by several countries over the course of its history but is nowadays recognised as Vilnius, the capital city of Lithuania.

He had studied Civil Engineering in Germany while also becoming a world-class tennis player, competing in the German Davis Cup team from 1928. In 1927 Prenn became the No.1 tennis player in Germany, a position he held until 1932, the year in which he beat Britain's Fred Perry in the Davis Cup.

Coming from a Jewish family, he chose to exile himself from Germany following the rise of National Socialism. He fled to Britain in 1933, and was granted papers (a very rare occurrence) in 1940. Daniel Prenn was clearly a shrewd businessman, and on his arrival in the UK he immediately set about immersing himself in British commerce. Over the next 50 years he created a large corporate empire, beginning with electronics and communications and expanding to other diverse areas, from floor cleaners to underwear.

Prenn's first business in Britain was assembling fountain pens, while his Truvox Engineering business was formed in June 1935 in Kentish Town (North London), manufacturing sound reproduction equipment, including one of the earliest portable reel-to-reel tape recorders—

Opposite: Truvox/Rola Celestion trade fair stand, early 1950s.

Truvox also produced one of the earliest portable reel-to-reel tape recorders—an area in which they were to become highly respected as pioneers

an area in which they were pioneers—as well as re-entrant public address speakers and miniature 3in loudspeaker units, although how Prenn made the leap from writing instruments to sound reproduction remains unclear.

By 1937 Truvox had settled into the empty Wembley Exhibition Grounds, from where it produced horn speakers, PA systems and radio sets as well as thousands of artillery loudspeaker-telephone sets for the forces during the Second World War. The company's portfolio covered the whole spectrum of the audio market, including loudspeakers for cinemas. Its subsequent range of public address systems was a legacy of the many acoustic devices that were developed for the Armed Forces between 1939 and 1945. Truvox's few post-war enclosed speakers included the 'Model 55' extension speaker, intended for radios or office dictaphones.

But it didn't stop with audio. The company also made a range of modern domestic appliances, including irons, kettles and desk lamps, as well as commercial floor-cleaning machines. To say the product pipeline was immense is probably an understatement.

Coincidentally, around the time of the acquisition, a number of alternative product lines were being attempted at Rola Celestion, including cigarette cases, cuckoo clocks and perhaps most strangely of all a pre-Christmas production of toy ducks manufactured on the cone felting machine! A more logical diversion was that of a moving coil microphone, although that too came to nought. Loudspeaker production at this time was purportedly around 35,000 units per week.

In the November 1953 edition of *Wireless World,* it was announced that the sale, distribution and service of Rola and Celestion loudspeakers, along with Truvox public address loudspeakers, would 'henceforth be undertaken by Rola Celestion, Ltd.' and not, as previously, by Truvox. Although trading as Rola Celestion, the company continued to manufacture products branded as Rola, Celestion and Truvox. However, by the late 1950s, it was Celestion that prevailed as the dominant brand name.

Truvox immediately brought a greater range of public address loudspeakers into the combined range, and as the dust of the takeover settled, Rola Celestion began manufacturing this new wave of public address components: pressure drive units, re-entrant and exponential horns,

Daniel Prenn (*second left*) and Les Ward (*third left*) with Truvox display their wares at a trade fair, *c.* 1951.

alongside the mains-powered and permanent-magnet speakers and transformers for which the brands had been noted throughout the 1930s and 40s.

Around this time, Jelinek the hairbrush manufacturer moved out of Ferry Works, and Rola Celestion wasted no time in occupying the whole building, which gave the company the space required to manufacture the additional products in its newly expanded portfolio.

A Truvox Sound Reproduction Equipment catalogue—dated August 1954—with the strapline 'Manufactured by Rola Celestion' contained the types of products you would expect from that period: large and small re-entrant speakers—with and without transformer—but also a dedicated range of compact reflex re-entrant speakers (RA13, RB13, RC13 and RA13/T) described as 'modern counterparts of Truvox designs which were proved by exacting wartime service in tanks and warships, on guns in action, in canteens and garrison theatres, and in

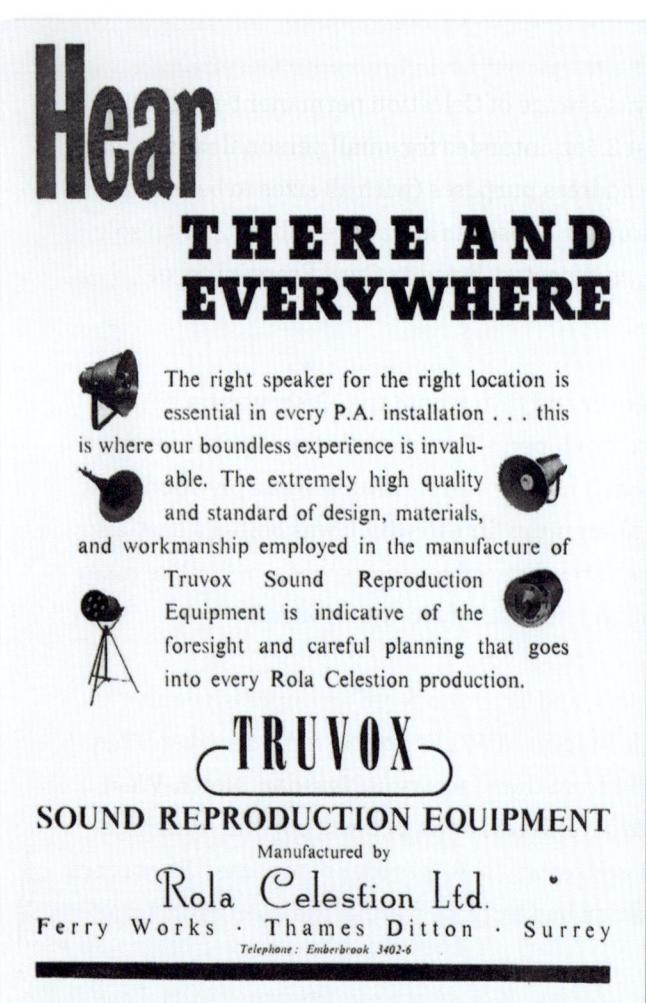
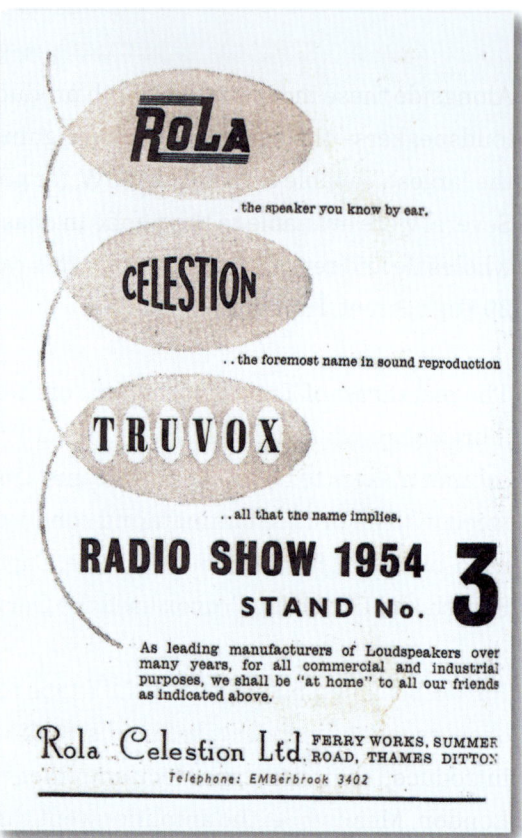

Left: Early-1950s advertising.

Right: Advertisement in the directory for the Radio Show, 1954.

general mobile work'. These were designed 'primarily for use where clear reproduction of speech is of paramount importance'. The diaphragm and voice coil assembly were moulded in fabric-reinforced phenol formaldehyde, ensuring long life under extreme stress.

The company also produced a range of pressure-type loudspeakers with magnet, diaphragm assembly and optional transformer, enclosed in aluminium case for horn loading. Then there were 46in-long exponential horns (with 24in flare diameter) and 54in reflex horns, as well as a tripod-mounted, 360°-rotatable 'Super-Power' long-throw speaker (the LH/100), which, according to the blurb, proved 'highly effective for propaganda purposes in war zones'. This contained six drive units coupled to separate re-entrant sound columns.

Bob Smith, who joined the company in 1968 as a 25-year-old development engineer, recalled working with highly influential chief engineer Les Ward, who had arrived at the company via Truvox. 'We were still producing pressure drive units and a 40in large-mouth horn for racetracks, et cetera, and smaller ones, re-entrant horns and flame proof loudspeakers for "hostile environments".' All had become mainstays of the range in the early 1950s and were designed by Ward.

Alongside these industrial-strength products was a range of Celestion permanent-magnet loudspeakers—the smallest model weighing just 3.5oz, intended for small personal radios, and the largest, capable of handling 40W, for public address purposes (with all sizes in between). Several were available to the public in chassis form or housed in attractive cabinets, with sole wholesale and retail distributors for this range, as expected, listed as Cyril French Ltd, of 29 High Street, Hampton Wick.

The real thrust of Truvox at the time of the takeover and throughout the 1950s was tape. A breakthrough had come in 1951, when Truvox developed a 7in domestic tape deck intended for cabinet makers to install in radiograms. Uniquely, it used a $^3/_8$in furniture-grade plywood deck, faced with a Formica laminate and conductive aluminium film, to which was bolted a cast-alloy head-block and flywheel/motor cradle. Capstan sleeves determined tape speeds, but unlike many British and US decks, Truvox used the German right-to-left, upper-track format.

It soon became a major force in DIY tape recorders, and Truvox offered an impressive range of accessories, including its famous 'Radio-Jack' plug-in MW/LW crystal radio receiver. They introduced their first tape-deck amplifier in 1954 from their factory at Neasden, North-West London. Meanwhile, the amplifier went through several revisions, proving popular with high-street HiFi dealers for the DIY enthusiasts and even other tape-recorder companies. Recorders even began to be marketed under the Rola Celestion badge; these were of the 'Recordon' type.

At the beginning of 1956, the *APAE Journal* and *Wireless World* both reported that 'W.H. Page and S.J. Tyrell, who joined Celestion in 1929 and were Directors from 1946, have now been appointed Joint Managing Directors of Rola Celestion Ltd. Mr. Page ("Billy") will continue to control sales, and Mr. Tyrrell ("Jim") design and manufacture'.

In 1957 Rola Celestion exhibited at the Audio Fair, which took place in April at the Waldorf Hotel in London. In the same year, the company announced the '415', employing two pressure-driven units to cover audio frequencies from 350c/s upwards, the lower-frequency unit driving a re-entrant horn, and by 1958 a new addition to the range for the high-quality audio enthusiast was the 15in coaxial 'Colaudio' speaker.

The Colaudio was designed to create a new realism in reproduced sound. It combined a 15in direct bass radiator with a 3in voice coil, and two direct radiator pressure-type high-frequency (HF) reproducers mounted centrally, in column form, inside the woofer, thereby radiating both low-frequency (LF) and HF signals along the same axis. Intended for the DIY enthusiast, it was said to be 'the culmination of over 30 years' R&D [research and development] in the search for an all-purpose solution'. Advertising implored listeners to apply the 'aural tone test', arguing, 'once heard, you will never be satisfied until you instal one in your own reproducing equipment'.

Further innovations followed. Celestion also tackled the problems of inhibiting cone break-up and relieving stresses at the corners in rectangular 'slot' speakers in its 2in x 2¾in C28 model by means of two long strengthening 'bulges' in the cone covering most of the cone length on either side of the voice coil, and by making the rolls in the surround larger at the corners and across the breadth than along the length.

Bob Smith also remembered Celestion had produced a number of speakers for British radio and television manufacturers—mainly 4in and 5in, as well as some smaller 3in transistor radio speakers. These were supplied to the heavy hitters of the day, including London-based Roberts Radio and Bush Radio as well as the even older EK Cole (EKCO), based in Southend, and US-owned British TV and radio manufacturer, Kolster Brandes.

The arrival of the stereophonic long-playing record by the end of that decade would have a significant impact, as enthusiasts of high-fidelity sound were keen to build and improve their own equipment. In order to take advantage of this, Celestion produced the G44/1300, which enabled an existing radio gramophone to be modified to incorporate two 12in G44 speakers (a variant of Rola's popular and versatile G12 format moving coil speaker) and HF1300 high-frequency units, which would soon take on a starring role when the company began building its own complete HiFi speakers (more on this later). Likewise, a special enclosure could be constructed to house these units, and for many this was their first experience of stereo sound. The system retailed for the modest sum of £18.10s, and this included the speakers, the enclosure design and full installation details.

New approaches were also taken by the Celestion R&D staff to perfect the performance of the coaxial speaker models, and two new types, the CX1512 (15W) and CX2012 (20W), with a much larger magnet, were introduced in 1962. These were both 12in PM units and, in common with Colaudio, had an HF unit built into the centre of the bass cone and designed with a 'pure electrical crossover' (i.e. it was deemed that no external circuitry would be needed).

In particular, it was the HF solution that represented an improvement over the earlier Colaudio, with the CX1512 featuring a variant of the almost ubiquitous HF1300 direct radiating treble unit and the CX2012 a compression horn driver, whose horn was built 'for wide angle distribution of high frequencies'. The CX2012 was also notable for the 'brilliance controller', enabling the listener to adjust the amount of treble to suit room conditions.

This story is typical of Rola and Celestion during this time, which, in terms of its commercial outlook and product range, could be considered an engineering-driven designer of component parts. These parts were the fundamental sound-producing elements that ended up inside a television, a bespoke public address system or a domestic loudspeaker cabinet, all built by one of the big names of the time with somebody else's branding on the outside.

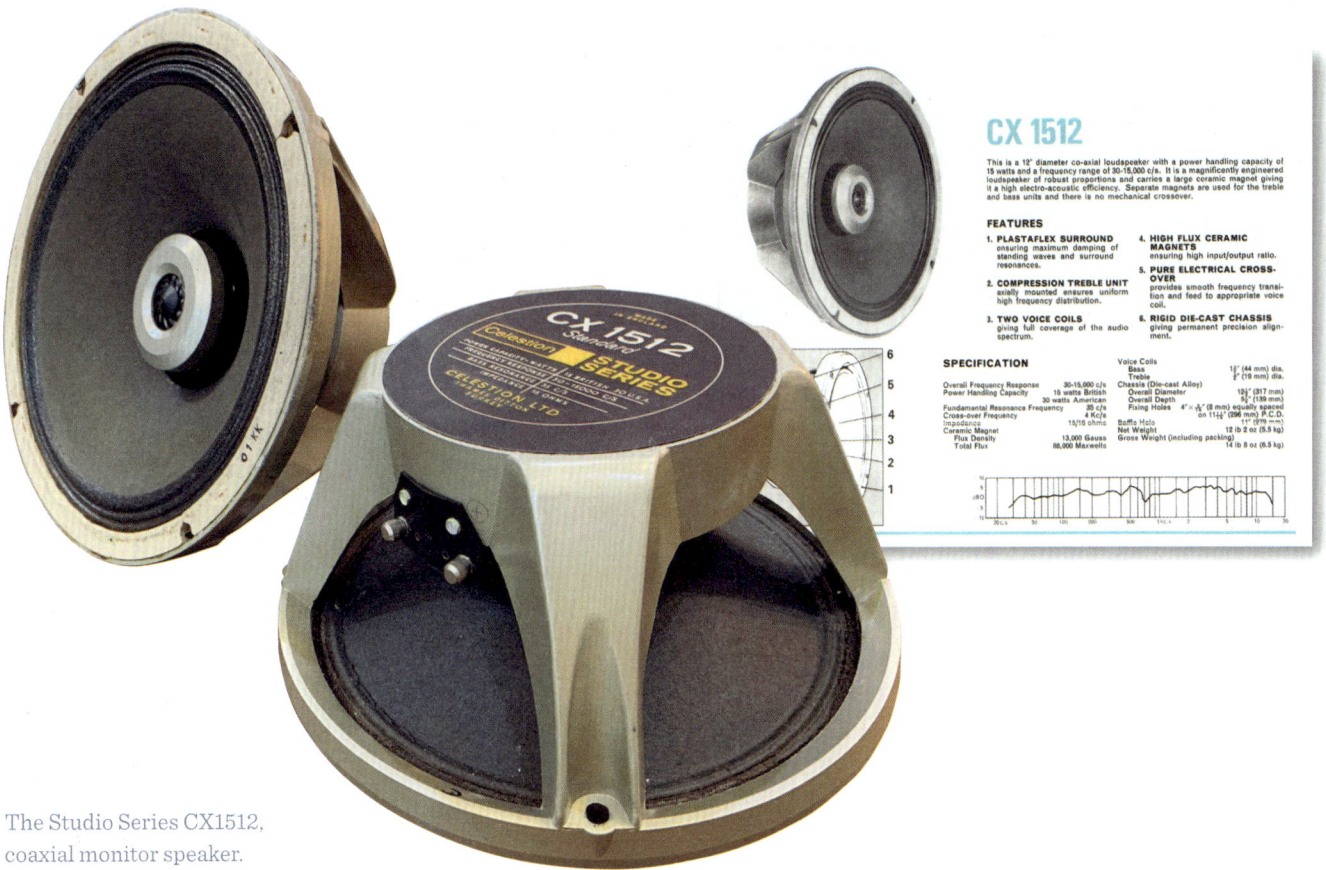

The Studio Series CX1512, coaxial monitor speaker.

However, with so-called HiFi becoming ever more popular as a hobby and leisure pastime, it was in this arena that Celestion's commercial approach was about to change, from component supplier to home entertainment equipment retailer once more, just as it had been in the earliest days of the company. In 1964 the Celestion engineers devised the first of many notable Celestion HiFi cabinet designs, satisfying a requirement that—since the commercial advent of stereo music reproduction—had yet to be truly addressed: affordable, quality HiFi speakers for the home.

The gradual implementation and eventual success of the Celestion HiFi range was the second of two turning points that were destined to transform the company. Both had their roots in the late 1950s, not more than a handful of years after the company had finally found its identity once more following the takeover by Truvox and the subsequent assimilation of that company's people and products under the Rola Celestion banner.

The first event was the ingenious adaptation of another—now venerable—loudspeaker component, the 'General Purpose' G12. This particular speaker, which had started life as Rola's foremost radiogram transducer, had previously been modified and updated for various applications, and the G12 format was thus available with different impedances and magnet strengths and even as the GL12, with a rubber surround (presumably for greater bass response). However, it was Rola Celestion's chief engineer, Les Ward, himself a former Truvox man, who adapted the format still further, meeting the demand for louder, higher-powered speakers and, in the process, initiating a legend that endures to this day.

TRUVOX AND THE EMPIRE OF DANIEL PRENN

6 Vox, Marshall and the Evolution of the G12

The story of the G12's evolution is a tale of smart engineering know-how together with a pioneering and collaborative spirit, combining to create a new kind of product that became more than just the sum of its parts. A loudspeaker that, along with two amplifier companies who went on to become instantly recognisable names, made an indelible mark on the sound of a decade and has continued to impact music making ever since.

Jennings Musical Instruments

It was the arrival of skiffle around the mid-1950s that created the first stirrings among the emerging musical instrument shops around London's Charing Cross Road. A confluence of jazz, work songs of the American Deep South and jug-band music, founded on folk song roots, it rapidly became popular on the British scene, before transforming into the harder-edged sound of rock 'n' roll.

At this point, the need for bands to 'amp up' became palpable, as the barriers between audience and artists were toppled. Redeployed cinema venues, some on their last legs, took on a new lease of life and became filled with noise, as 1950s rock 'n' roll 'package tours' started to populate these venues and fans, eager for a glimpse of their new heroes, were filled with excitement.

Acoustic guitarists of all stripes had already started beating a path to the hugely influential and eclectic Mairants Musicentre, in Rathbone Place, London. From 1958 they had started selling a range of guitars that attracted jazz and folk blues specialists alike. Around the corner, the Jennings shop at 100 Charing Cross Road had specialised in accordions and the organ, but with the coming of this new musical revolution it was soon to change.

The first company to recognise the growing need for more amplifier power was another Charing Cross Road resident, Selmer. However, it was Jennings that found itself at the vortex of skiffle, and owner Tom Jennings immediately saw an opportunity to grow his business with a guitar amplifier—not least because skiffle pioneers such as Lonnie Donegan were replacing the washboard and Spanish guitar with drum kit and electric guitar, which now required some kind of amplification for the vocals.

Opposite: A B024, 8Ω alnico speaker loaded in a pre-1960 Vox AC15.

The Jennings shop on Charing Cross Road.

At around the same time, Dick Denney—an electronics wizard—was recovering from a serious illness and, while recuperating, he came up with the idea for a compact guitar amplifier for use with his Hawaiian guitar. Denney was a big band guitarist also, and rumour had it that this was largely born out of the necessity to hear himself play, since he was going deaf.

By good fortune, he visited Jennings' music shop to show him the fruits of his labour: a 15W guitar amp he had just designed and built. Jennings hired Denney on the spot, and in 1956 Jennings Musical Industries (JMI) was born.

This 15W amp would become JMI's first guitar amplifier. Marketed under Jennings' Vox trademark, it would quickly lead to the highly regarded AC15 (originally designated the AC1/15), which was an improved, second-generation version designed in 1958. This was also the year JMI would develop and release the iconic Vox AC30, with its four cathode-biased Mullard EL-84 output tubes, which later fuelled the so-called 'British invasion' and was thereafter embraced by rock 'n' roll royalty.

As the decibel levels created by screaming fans became ever higher, it was the Shadows' guitarist, Hank Marvin, who had pleaded for something more muscular than the AC15 to enable his signature twang to be heard at their shows, backing teen idol Cliff Richard. More than anything else, it was this need that triggered the development of the AC30.

Dick Denney came up with the idea for a compact guitar amplifier for use with his Hawaiian guitar

The original AC15 had used Goodmans Audiom 60 speakers, rated at 15W peak, but around 1959 the switch to Celestion had begun, using a modern variant of the original Rola G12 radiogram speaker first built in 1936.

Vintage Vox enthusiasts have since noted that the versions used by JMI in earlier amps were models designated B024 (8Ω) and B025 (15Ω). These were likely adaptations of the Celestion C64 radiogram speaker—which, along with the P44, was a contemporary permutation of the G12 template—that had been in production through much of the 1950s. Featuring a motor built with an alnico magnet (alnico being an alloy of aluminium, nickel and cobalt), both speakers were painted silver and issued with a standard Rola or Celestion label: at this time, the two brands were almost interchangeable, and the same product might often be labelled with either company name.

The single-speaker AC30 of 1959–1960 had generally been fitted with the 8Ω B024. In the dual speaker AC30/4s (with four inputs), it's possible to find pairs of 15Ω B025 wired in parallel, or of 8Ω B024 wired in series. Also found in AC15s at around the same time was the CT3757. Although similar in construction to the B024 speaker, it featured a Vox-branded rear label, and examples have been found dating up to 1962.

Few details about the CT3757 exist; however, it's thought that it was a transition model, possibly a prototype or a stepping-stone to the much more famous T530 which was to come. In fact, the entry in the Celestion 'T-book' for the T530 reads: 'G12: As CT3757—sprayed colour No.M104 azure blue—fitted with cover H1668.'

The AC30's output demanded more than the 12W speaker power handling that the B024 and B025 could offer. So JMI's Derek Underdown and his right-hand man Alan Harding worked with Celestion Chief Engineer Les Ward, seeking ways to strengthen the G12's construction and give it greater power-handling capacity for this new application.

Interviews at the time recorded that the Vox designers 'didn't want to use negative feedback (in the AC30 circuit); they wanted it to go into resonance'. So, in order to withstand that as well as all the additional rigours of being a speaker in a guitar amplifier, first the G12 cone was 'doped' (a treatment to the edge of the speaker cone which affects the sound and acts as a shock absorber). In this case, a clear material was hand-painted on ½in of the edge of the cone to strengthen the surround—the part most vulnerable to mechanical breakdown; this is common in contemporary

Cliff Richard and the Shadows; Hank, Bruce and Jet play through their AC30s.

speakers but was a practically unknown process at the time. The doping also restricted the cone's movement, which kept the voice coil better centred in the magnetic field, resulting in increased efficiency and reliability.

The second modification was to change the windings of the voice coil from aluminium to copper, which would remain stable to higher temperatures. This helped keep the coils from melting due to the heat generated within the magnet motor when the speaker was driven hard by the amplifier. Finally, the termination wires were made stronger, all the better to cope with the fast, high amplitude movement that the amp demanded.

By August 1960 the new prototypes based on these adaptations had been accepted. Once officially approved, the new speaker was assigned a T number, and special Vox labels were designed for the magnet backs, a design also used on many other Vox speakers for years to come.

Rated at a more robust 15W, this first iteration of the T530 was initially produced in Hammertone/Oyster paint finish (a slightly pinkish-silver colour), but by January 1961 it was supplied in azure blue at the behest of Tom Jennings, who was thought to have favoured the colour. Thus, the legendary 'Vox Blue' was born.

Left: Entry in the T-book for T0530.

Right: T0530, the Vox Blue.

The principal difference from the first few 'oyster' T530s (aside from the colour) was that the Vox Blue also had an additional bell-shaped aluminium cover fitted over the magnet. This didn't affect the performance of the speaker, but it did help shield the magnet from foot-pedals and cables rattling around in the amplifier case. The Blue was made available in an 8Ω version and a 15Ω version (designated T727). It quickly became popular in the AC30 Twin and was later used in the AC15.

> **'T numbers'** had become the prevalent method of part numbering from the early 1950s onwards, superseding all other numbering systems that had been used previously, although CT-numbered parts were built right up until the early 1970s. The 'T-book' became the definitive catalogue of Celestion part numbers, and continues to be so to the present day.

Legend also has it that Les Ward had spent some time with Tom Jennings experimenting with speaker and cabinet configurations. They found that an open-backed cabinet together with the new, tougher G12 gave a louder, less directional sound. 'For the first time the sound flooded the stage,' a quote attributed to Ward years later recalled. 'Before this, the guitarist didn't have the volume for clean, undistorted playing at the levels required for larger venues.'

Once Hank, Bruce and Jet strode out with their AC30s performing as the Shadows behind Cliff Richard, the amp immediately fulfilled its objective of overriding the screams of their audiences and became the go-to brand of the time. Its unique tonality immediately found favour with The Beatles, The Rolling Stones and other trailblazers. And, as volume levels intensified, this in turn fuelled the requirement for more substantial 50W and 100W valve amp circuits.

Released in 1966 the UL730 was used extensively by the Beatles during the recording of *Revolver*, and to a lesser extent, *Sgt. Pepper's Lonely Hearts Club Band*

Vox was given exclusive use of the new T530. In the end, the lifespan of the product was a little under three and a half years, but in that time it achieved near-mythical status, ultimately becoming the most beloved of all the Vox speakers. In 1964 there was a year-long cosmetic change in which all Vox amps were converted to a monochromatic colour scheme. Celestion was folded into this change, and in 1964 the familiar azure blue had all but disappeared, retired in favour of the new colour—'poly grey'—with the speaker itself becoming a new product, designated T1088.

Also introduced in 1964, the first AC50s did little more than the AC30 to make themselves heard above the din of the audience, and so JMI responded by doubling both the wattage of the amp and number of speakers in the cabinet to produce the Vox AC100 (80W–100W) amplifier. The later MkII version (AC-100/2) corrected design flaws in the original, with a new Class AB amplifier output. Celestion's T1088s, with light doping, were employed in the AC100 cabinets used by John Lennon and George Harrison in 1964.

Later the UL730 was introduced as a solid-state replacement for the AC30 valve amp. Released in 1966, according to anecdotal evidence, it was used extensively by The Beatles during the recording of *Revolver* and, to a lesser extent, *Sgt. Pepper's Lonely Hearts Club Band*. The associated open-backed speaker cabinet contained two of the alnico speakers.

Well over 50 different transducers have been used in Vox amplifiers over the years, including the G12M Greenback, which was to become the second iconic Celestion speaker in the latter part of the 1960s, and the two brands still enjoy a strong relationship to this day. However, it was the pairing of the unique electronics of the AC30, together with the specially modified alnico G12, that was for many the magic sound of Vox. Moreover, it was the development of the T530 'Alnico Blue' that set in train a whole portfolio of G12 derivatives.

Marshall Amplification
In July 1960, just as the G12 T530 was moving into full production, drum teacher Jim Marshall opened his first music shop with wife Violet and son Terry on Uxbridge Road, Hanwell, just a few miles west of Central London.

They started with drum equipment, as Jim Marshall told it, but 'then the drummers brought the rest of their groups in, including [players like] Pete Townshend, and said "why don't you stock guitars and amplifiers?"' Marshall duly did this, and it went down a storm with the musicians who frequented the place, sometimes just to hang out.

However, all of the amplifiers were imported from the United States, and, as a result, were both limited in number and expensive. There was also an increasing clamour for a different-sounding, louder amp. The US sound was considered a little 'too clean' for the British guitarists, who seemed to be more interested in distortion and lots of it, and, even more importantly, they just weren't loud enough.

Terry Marshall, who was working part-time at the shop while also gigging as a saxophone player, already knew many of the players from the local live music venues and convinced his dad that it would be a good idea to try and build an amplifier of their own. Terry had no electronics experience, but was convinced he knew the sound that the local guitar players were looking for. He sat down with Ken Bran—a guitar player who had often visited the shop along with his own band until he had been recruited by Jim Marshall as a service engineer—and electronics engineer Dudley Craven, in an attempt to put together an amplifier.

They began with an amplifier built by Craven and based on a tried and tested RCA-designed circuit which hadn't been subject to trademark so was free to use. It was a popular circuit at the time and likely the same one on which the Fender Bassman amp had been based. However, Marshall wanted to upgrade the power stage for a beefier output, and to modify the preamp in order to find the 'right' tone. Terry Marshall played through the prototype amplifier (using both a Fender Stratocaster and a Gibson 335), with Ken and Dudley changing components at his behest, until they found the sound that Terry believed his guitar-playing friends had been looking for. 'As a saxophone player I often shared the stage with three or four guitarists,' he explained, 'and from hearing the sounds they were trying to achieve it helped us to get from the amp the sound they were looking for.'

The result was the JTM45 (JTM being short for Jim and Terry Marshall). It had a noticeably more aggressive sound and delivered a true 45W RMS output power. Many orders were taken on the back of this completed prototype, with the very first production model being sold to Pete Townshend, who had been desperately seeking an amp that would deliver more power.

When it came to speakers, the Goodmans Axiom speaker had been tried but deemed unsuitable. 'They couldn't handle the power, they kept bottoming out,' recalled Terry Marshall, and a subsequent trip to the Fane factory in Yorkshire had proven unsuccessful. Meanwhile, Ken Bran had already used the Vox G12 in amps he had built for himself, so it seemed like the natural next

Left: The original Marshall 4x12 cabinet.

Right: A Marshall-labelled T0652.

step for Marshall to try them with the JTM amp. The sound of the G12s passed the listening test with flying colours, and almost by accident a partnership with Celestion was initiated which lasts to this day.

The version of the G12 that bore the T530 nomenclature had been made exclusively for Vox; instead, Marshall used the Celestion-branded equivalent, which became known by the part numbers T650 (8Ω) and T652 (15Ω). Constructed with the same alnico magnet assembly as T530, these models were sonically equivalent to the speakers built for Vox but cosmetically different, sprayed silver and supplied without the magnet cover.

Originally, they had featured a standard Celestion label fixed directly to the back of the magnet, but, as the partnership with Marshall solidified, this was soon switched to attractive custom labels instead. These labels are also a good indication of the manufacturing date for early Marshall cabinet models, specifying either Uxbridge Road, Hanwell (1962–1964): the location of the two premises where the early amps (in the back yard of number 76) and cabs (in the back yard of number 93) were built; or Silverdale Road (1964–1966): the first dedicated Marshall factory location in Hayes, Middlesex.

Although the G12 guitar speakers had originally been toughened to withstand the greater output of the 30W AC30, the 45W Marshall amp was ultimately too much power for two G12s to withstand. Jim Marshall had a novel idea to fix the problem, and the very first 4x12 cabinet design was conceived.

Left: The Marshall 1962 'Bluesbreaker' 2x12.

Right: The legendary 'Beano' album cover.

In the early days, 'Marshall production was very much hand to mouth. We would sell one amp and that would pay for the production of the next one,' remembered Terry Marshall—but demand quickly escalated. 'And, of course Rola Celestion were growing, the same as we were, so every now and then I had to go to Thames Ditton myself and pick the speakers up in an estate car': so rapid was the rise in popularity of the new G12s that production at Celestion sometimes struggled to keep pace with demand. 'The factory was a lovely old building in a beautiful setting by the River Thames. I had a close relationship with the Rola guys and occasionally we'd go to the pub next door [The Old Swan] for a drink.'

Largely to accommodate Marshall's newly developed 100W amps, which were first built in 1965, Jim Marshall had built a small number of 8x12 cabinets, purportedly at Pete Townshend's request. Townshend himself and The Who bassist John Entwistle had each bought two of the monster cabs all loaded with alnico G12s.

A further two 8x12s had been built at the same time, eventually to be bought by fellow Londoners the Small Faces. Soon after this, however, the format was deemed too heavy and unwieldy to carry, particularly as the cabs had been fitted with highly impractical bar handles. Instead, Jim Marshall instigated the building of two separate 4x12 cabinets intended to be placed one on top of the other, and just like that the modern speaker stack was born.

The amp that was affectionately dubbed 'Bluesbreaker' by Eric Clapton in reference both to his band at the time, John Mayall & the Bluesbreakers, and the associated 1966 album *Blues*

Breakers with Eric Clapton, was in fact the Model 1962 a 2x12 combo which, at least to begin with, used 15Ω silver alnico speakers. Terry Marshall explained, 'It was the first 2x12 we made. It used the same amp head as the JTM45, but combined with two 12s.' Not only was the Bluesbreaker Marshall's first combo amp, but it became the template for a signature sound that defined British 'blues rock' in the mid-1960s.

Affordable, practical permanent magnets were still a relatively new commodity and, like nearly all loudspeakers up until this time, the G12s featured an alnico magnet. By the mid-1960s the rare earth metals that constituted alnico were becoming increasingly expensive, principally due to the locations where they were found and mined. Rhodesia—an important source for nickel—was in the midst of a civil war (which ultimately led to its independence as Zimbabwe in 1979), while cobalt came from the Congo (now the Democratic Republic of Congo, which had been at war since 1960) or from Soviet Russia, making access difficult.

An iron-based ceramic was found to be a suitable alternative to alnico: it was similar in weight and magnetic strength (though not in shape), with the advantage that these ferrite magnets were much cheaper, the raw material being more easily obtainable. They were duly adopted and began to supplant alnico as the preferred, cost-effective choice, and by 1965 the ceramic G12M, part numbers T1220 (8Ω) and T1221 (15Ω), came along. They were described in the T-book as having a 'plastic can', which turned out to be an olive-green cup that fitted over the magnet assembly, more than likely to cover the ring of ceramic material, which was brittle and tended to chip if not handled carefully.

These new G12 speakers had also been constructed slightly differently from the original alnicos, with kraft paper being used for the voice coil formers in preference to cartridge paper. This modification, along with the greater surface area of the ceramic magnet assemblies, afforded the speakers a little more power handling. The first G12Ms were rated at 20W (and revised to 25W a couple of years later) making them more robust and leaving a little more 'headroom' when used with a higher-power amplifier.

Affectionately referred to as 'Greenbacks' (because of the rear-mounted can's colour) the G12Ms quickly caught the imagination of Marshall and its growing legion of soon-to-be-legendary players—from one James Marshall 'Jimi' Hendrix and Eric Clapton to subsequent Yardbirds axemen Jimmy Page and Jeff Beck, and Free's Paul Kossoff to name a select few—quickly replacing the alnico G12 to become *the* standard speaker.

Opposite: Jimi Hendrix plays in front of at least two full Celestion-loaded Marshall stacks!

This was the era guitarists started to push amps hard to distortion, reaching for greater volume and driving their amplifiers more and more into overload

Ultimately, it was the unique sound of a G12M paired with Marshall's valve amp just on the edge of distortion—characterised by a warm, controlled low end, a rich, vocal mid-range and a delicate, detailed top end—that assured its success. It is notably manifest on Jimi Hendrix's early recordings and can be heard on countless recordings from that time onwards, as the story of Marshall amps and the Celestion Greenback quickly became intertwined.

The G12 Evolves
The G12M, however, was not the first G12 to be built with a ceramic magnet. That accolade went to T1134, described as G12H and available in 10Ω and 12Ω impedances (for completeness, T1161 was the same model 'with a standard 15Ω coil'). Elsewhere, T1164, T1174 and T1175 were all listed as G12C, each with a subtly different parts specification and all appearing slightly before the G12M.

By now a nomenclature for the ceramic magnet speakers was emerging, with a letter designating the size or at least the type of magnet. M referred to a 35oz 'Medium' magnet and H to a 50oz 'Heavy' magnet, with the G12L 20oz 'Light' magnet appearing around 1966. Somewhat opaquely, the C magnet simply denoted 'Ceramic' and carried no information on weight. And as the number of G12 speakers proliferated into the 1970s and beyond, so too did the magnet codes.

As the pursuit for greater power handling began, the G12H fully came into being, reinvented with its own green can. The heavy 50oz magnet initially conferred a 25W power rating on the speaker (soon after, this was uprated to 30W). It was the G12H variant, with the lower resonance, 55 cycles/sec, bass cone: T1234 (8Ω), T1281 (15Ω), that was adopted by Marshall in a cabinet intended for use by bass guitarists but which proved popular with lead players and was quickly adopted for that purpose.

This again saw Hendrix at the forefront of a trend, this time using the 4x12 G12H-loaded bass cabinets during live performances, for the extra power handling, and also in some of his later recordings, perhaps because of their additional heft and the warm, 'syrupy' tones they delivered.

Reflecting on the genesis of these developments, Ed Form, chief engineer at Celestion between 1981 and 1987, explained the rationale: 'Les Ward had taken [the G12] that was designed to go into console radios, beefed the power up [for use by] Vox and Marshall and they took off because of both their power handing and tonality. This was the era guitarists started to push amps hard to distortion, reaching for greater volume and driving their amplifiers more and more into overload as they strove for something even louder with a more sustained sound.

'But when a valve amp is pushed into overload it produces increasing amounts of odd-order distortion [3rd and 5th harmonics] and under extreme overload, when the electron cloud in the valve can be seen as a blue corona, this can be a big problem, sonically speaking.'

It was a complete disconnect from what was taking place at the same time in the United States. According to Form, the US idiom was sonically less aggressive. As such, the unwanted 3rd and 5th harmonics weren't being generated by the amplifiers; it was an altogether much cleaner sound. 'If you were to put, say, an American speaker like an Electrovoice or JBL into a UK cabinet it sounded unpleasant as it fairly accurately replicated the distortion sound in all its angry glory.'

Overdriven valve amps using US loudspeakers tended to reproduce that more aggressive output, and consequently didn't sound so good. But Les Ward's speakers had a spectacularly different behaviour compared with the US drivers.

According to Form, it was down to a clever piece of design: 'Ward discovered that you could pass the coil through a roller to produce a groove in it, and then wind it round the former; this created a compliance, resulting in a loss of accuracy but more importantly in the softening of the distortion. This was a key part of the Celestion G12 guitar speaker behaviour: a gorgeous smoothing of the overloaded sound which worked in concert with that famous Marshall saturated distortion and of course so many other amplifiers afterwards.'

This totally new sound became the characteristic distorted guitar sustain of British rock music. 'For one example of this,' explained Form, 'listen to Paul Kossoff of Free playing one of his solos; he seemed to be able to sustain for ever but his sound, though hugely distorted by the amplifier, was infinitely sweet – that's what the combination of an optimised valve guitar amplifier and a bank of Celestion G12 loudspeakers made possible!'

And, crucially, while producing the desired tonality, they were able to withstand the output of the new valve amplifier circuits. 'The tendency of the Celestion G12s to be inaccurate was good fortune and that was how the British rock 'n' roll sound came through.'

7 Big Sounds, Big Names

If Vox and Marshall were the guitar amp brands most famously associated with Celestion from the beginnings of the rock 'n' roll scene in Britain, then other well-known brands, including Selmer, Carlsbro, Simms Watts and more, were also loading their speaker cabs with Celestion chassis as the 1960s progressed, in an attempt to establish their signature sound on the touring circuit. Loading their amps with Celestion G12 speakers would give them real gravitas as they sought to create *that* tone.

The years 1967 and 1968 were also critical: the G12M and G12H, now well established as the go-to guitar speakers, were updated to handle more power. It was also the time when two of the most prominent British amplifier start-up companies were Laney and Orange—both companies originated from musicians striving to reach the holy grail of tonal perfection and both had received their induction into Celestion's own form of alchemy while playing through Vox AC30s or Marshall 4x12 stacks.

Those years also defined the coming-of-age of Charlie Watkins' WEM (Watkins Electric Music), after delivering his famous 'Wall of Sound', the world's first 1000W sound system. As the versatile G12 was adapted for use as a public address (PA) speaker, Celestion slowly took over as WEM's engine for ever-louder PA, marking the beginnings of the company's journey into sound reinforcement.

Laney Amplification

Over the years since their inception and right up to the present day, Laney Amplification has used a wide range of Celestion loudspeakers, including lead guitar and bass products (for solid-state, hybrid, and tube amps). Set up in the West Midlands by Lyndon Laney in September 1967, immediately after he left school, Laney Amplification was soon operating at the hub of the burgeoning Birmingham music scene.

Opposite: Celestion-loaded WEM PA at the Isle of Wight Festival, 1970.

A very 'formal and old-school' gentleman, according to Laney, suddenly appeared at the door…"He said, 'it's not you I've come to see, it's your father'

As a bass player, Laney most notably played in Band of Joy, which would variously feature Robert Plant (vocals) and John Bonham (drums)—both later to achieve fame with Led Zeppelin—as well as Dave Pegg—later of Fairport Convention and Jethro Tull—in its line-up.

Laney started building an amp in his dad's garden shed and buying components, such as KT66 tubes from electrical retailer Radio Spares. The early-production amps had been born from necessity, he reminisced. 'Later, when I was asked what the reason was for starting to make amps, I replied "poverty". I simply couldn't afford to buy an amp, so I made one.'

'I used to buy what I could and made a 50W tube amp. I was then offered a sponsored place at Aston University to study electronics and electrical engineering, but never went, despite being accepted.' Long term, he knew the band life wasn't for him, and so he continued making amps. 'The band's manager thought it would be an idea to make a business out of manufacturing so I wrote to the university and said, "thanks but no thanks." I took the University of Life instead!'

It was when Laney was approached by (Birmingham-based music shop) Ringway Music that business really took off, and his attention turned to Celestion: 'Ringway said "it's nice amp but we need some speakers."'

Laney's first encounter with Celestion had been a few years earlier via a Marshall 4x12 that he'd bought in 1964. 'It contained four 15W alnico speakers and I think I had one of the first ones to be built. When I got home, I unscrewed the back and there were the Celestions.'

'Well, I knew nothing about speakers, but I started to learn and contacted Celestion in Thames Ditton. Soon after, we started buying a few G12M 25W speakers—the Greenback as it was known.' And, as demand for his amplifiers increased, the company moved from the back garden of Laney's family home into the iconic Bird's Custard Factory in Digbeth (in the heart of Birmingham) in September 1967.

Things accelerated even further when massive orders started to come in from a national retailer: 'This was a big deal for us. Suddenly we had to start ordering components and doing things properly.

Left: Lyndon Laney in the workshop c. 1967.

Right: Tony Iommi plays, in front of a wall of Laney amps and cabs.

'We appeared on the radar at Celestion and someone from the company wanted to come up and visit me, which was embarrassing, as our place was not exactly what you'd call salubrious. A very formal and old-school gentleman suddenly appeared at the door. I think his name was Page and there he was facing this 18-year-old scruffy kid. He said, 'it's not you I've come to see, it's your father'. I said, 'well I can't get him here until after 4pm as he is a teacher down at the school'. He had assumed my father was the owner of the business—and that was my first encounter with anyone from Celestion!'

However, the Greenback was constantly in high demand from the growing number of amplifier manufacturers using the speakers and, due to space constraints in Thames Ditton, production quantities were sometimes limited. According to Laney, 'Because we had an on/off situation, supply and delivery was sometimes a problem. Occasionally we were forced to buy from Fane and Goodmans, but it was not the same as having Celestion, which always sounded quite different. Frankly that magic is still there today, musicians know the way they sustain and the way they break up. Celestion is still prime for the sound they want. People would say, "I have a solid-state transistor amp that sounds like a tube amp"!'

One of the earliest devotees of Laney's amps was heavy metal pioneer Black Sabbath's Tony Iommi, who formed an enduring relationship with the company after gravitating to Laney's signature tone. From the earliest days of the band, Iommi used G12H speakers in his LA412 cabs: the heavy magnet speaker for the heaviest of sounds.

BIG SOUNDS, BIG NAMES

We painted the shop Orange—it was not only my favourite colour but also my favourite food

The Orange Music Electronic Company

Just a year after Laney, in 1968, Orange was officially founded by another bass guitarist (and electronics designer), Cliff Cooper. The company's HQ in New Compton Street on the edge of Soho was right in the heartland of London's Tin Pan Alley, a stone's throw from the heavily populated music shops of Denmark Street and Charing Cross Road.

Cooper had been playing bass guitar through a backline of Vox AC15s and AC30s with his band, The Millionaires, and so was well aware of Celestion. The band, which also featured Cliff's brother Ken, was one of the last recorded by British record producer Joe Meek, with the much-loved single *Wishing Well*. Reflecting on the company's origins, Cooper remembered, 'We painted the shop Orange—it was not only my favourite colour but also my favourite food! London County Council gave us a derelict shop. They were going to knock it down, but we ended up being there for 10 years.'

As far as guitar speakers were concerned, Celestion remained the primary target. 'We were well aware that the Alnico Blue, and the original Greenbacks made famous by the Vox AC30, were lovely speakers,' Cooper says. 'Celestion speakers had a unique and iconic distortion characteristic. The Beatles' *Paperback Writer* is a perfect example of that sound.'

Even the earliest Orange cabinets used either the Celestion G12H Greenback or an equivalent. 'The problem was that the other speaker brands we used didn't stand up to heavy use like the Celestions.' It was the G12H, along with its unique size and construction, that had helped Orange establish its name for great-sounding guitar cabinets.

Mick Dines, who was responsible for designing all Orange's cabinets over a 50-year period, also recalls the history: 'We first started using the G12H in 1968 and by the end of '69 the Celestion 30W Greenbacks were used exclusively in all our 4x12 and 2x12 cabinets and combos, although the demand for the speakers from all manufacturers always seemed to be greater than the supply!'

When Orange couldn't wait for delivery, Dines, as factory manager, would personally drive down to the Thames Ditton factory in Surrey and collect the despatch in his car—by now a tried and tested means for amplifier brands in a hurry to get hold of their speakers as quickly as possible. 'There would always be a wait early in the morning whilst their staff put the speakers in their cardboard boxes. They couldn't make them fast enough!'

The Orange shop in New Compton Street *c.* 1968.

He remembers crossing a small, shallow river ford to get to the aptly named Ferry Works on the bank of the River Thames—but on the return journey the water would be above the door threshold of the car due to the excessive weight of the speakers. When he got back to the Orange factory, his staff would be waiting to fit them in the cabinets and they would be shipped out later the same day.

Dines further recalls that production quantities of all models did increase when Celestion relocated to Ipswich, 'although we still carried on collecting by car when they were in short supply and we were desperate!'

Orange Technical Director Ade Emsley, a veteran of more than 25 years at the company and current designer of the Orange amplifiers, encapsulates the personality of Celestion in his definition of the unique harmonic distortion derived from the G12: 'This appears when saturation is reached in the output valves, which in turn saturates the output transformer. There is a unique synergy that occurs when a Celestion guitar speaker at the end of the amplifier chain is pushed in this way. What happens is the speaker brakes early in sympathy but always gradually, and never harshly. This is aided by the cone's unique suspension—which is a brilliant feature.'

BIG SOUNDS, BIG NAMES

WEM had been responsible for the first tectonic shift in festival sound reproduction with their famous 'Wall of Sound' slave PA

Watkins Electric Music

The relationship between Celestion and WEM, and their influence on the market during founder Charlie Watkins' pioneering days in the 1960s, cannot be overstated. WEM had been responsible for the first tectonic shift in festival sound reproduction with its famous 'Wall of Sound' system, providing 1000W power capacity for promoters Harold and Barbara Pendleton at the 1967 Windsor Jazz & Blues Festival. Barbara later recalled that this sound system 'hit the headlines in the local paper and frightened all the residents'.

The system was later upgraded further to 2500W for The Who at the 1969 Isle of Wight Festival, using up to 16 WEM slave columns at each side of the stage. It was announced at the time as the loudest sound system in the world, and the set carried warning signs for people to keep at least 15ft away!

Just a year later, it was the Isle of Wight Festival sound system that was the first to feature WEM's unique parabolic dishes for outdoor venues, using 10in Celestion speakers firing towards the dish, which in turn reflected back the mid-frequency (MF) and HF signals, sending them over a remarkably long distance.

WEM's long-serving wireman John Thompson, who, along with Norman Sergeant, had been a key player in WEM's early success, recalled in one interview, 'WEM 4×12 PA columns were used by hundreds of bands back then. They usually contained four Celestion G12 drivers.' He also remembers using Celestion MH1000 horn tweeters in a narrow column between two columns of woofers. In fact, WEM's tall, thin columns contained three MH1000 in each vertical horn box, with the four-way boxes containing four of these tweeters. That was the classic 4x12 public address column that WEM are best remembered for—the sound that took off: the thin column sandwiched by the two big ones per side was the classic 200W system.

The MH1000 horn, designated the T1360 by Celestion, was used by numerous companies, although Watkins specified them routinely, including to make wooden four-way sectoral horns, as well as the top end of WEM's 'Vendetta' PA columns. They were also used in a three-

Left: A single WEM PA stack.

Right: the MH1000 driver and horn combination.

way cabinet, built with two 12in and two 10in woofers and passively crossed over. For this configuration WEM built the horn drivers in vertical columns, and the MH1000 remained the order of the day until the mid-1970s, when its use was overtaken by the piezo horn.

In Celestion, Charlie Watkins had already found the perfect soft cones that moved easily, 'not like the horrible stiff things they were using for guitar speakers', Watkins later told the magazine *Sound on Sound*: proof, if it were needed, that the 'cloth edge' variants of the G12 speakers used for PA were significantly different from the 'standard' models used by guitarists.

Speaking of PA speaker development, Watkins added: 'Another thing I knew was that you needed a speaker with a decent hefty magnet, and I'd always had that with Goodmans and Celestion, not like those dreadful Jensen things they had in America—you couldn't get any balls out of them! I'd seen over the mountain with this new system, using Norman Sergeant's new amplifier and these soft-coned speakers in a proper box, and I could see the business was there for the taking.'

BIG SOUNDS, BIG NAMES

The WEM PA was upgraded to 2500W at the 1969 Isle of Wight festival.

Reflecting on his columns, he commented: 'Whatever you do with columns, you can't project bass from them, so I made an exponential horn—a gigantic one coupled to a box with four Celestions in it, facing each other into a 3in channel with the horn bolted on. Slade wouldn't play without them. That was a lovely bass sound.

'In fact, the man who ran the lab at Celestion was interested in what I was doing with this and asked me to take the system down there so they could test it in their anechoic chamber. So off I went and they put it in the chamber and measured it with their Brüel & Kjær gear, and afterwards they just looked at me and said, "How did you do that? It's dead flat!"'

Chris Hewitt of CH Vintage Audio faithfully recreated Pink Floyd's famous *Live at Pompeii* WEM PA set-up and described how it was organised. Each side (left and right stacks) comprised ten SL100 slave amps, two Audiomaster mixers, six WEM 4x12 columns with Celestion and Goodmans 12in speakers, two WEM 2x12 columns with Celestion 12in speakers, a WEM Festival Stack consisting of two 15in Celestion bass units, a 4x12 Celestion lower mid-range cabinet and a 6x10 Celestion upper mid-range cabinet with horn-loaded cone speakers. Additional horn-loaded pressure unit cabinets comprised a four-way Celestion MH1000 and a three-way vertical MH1000 horn box, plus additional Vitavox sector horns.

To make MH1000s work Charlie placed four vertically in a four-way box with wooden flares between the horns as used by Led Zeppelin, Fleetwood Mac and Pink Floyd

Hewitt explained, 'To make MH1000s work Watkins placed four vertically in a four-way box with wooden flares between the drivers.' These were used in sound systems by the likes of Led Zeppelin, Fleetwood Mac and Pink Floyd. Another variation saw two horn drivers placed vertically, either side, and two horizontally in the middle again with wooden flares; there was also a three-MH1000 column the same height as a 4x12 and 6in wide. Most of the 12in speakers used in this set-up were twin cone versions of the G12, including G12M Creambacks (a variant of the G12M produced in the early to mid-1970s with a cream-coloured can rather than a green one).

Another first for Charlie Watkins had been his pioneering use of transistor amps (having previously been possibly the first to deploy the EL84 valve). However, solid-state brought its own challenges, according to Hewitt: 'The industry had previously been using valves but early transistor amps that Charlie designed couldn't be run down to 4Ω. The lowest impedance they would go down to without overheating was 6Ω, so WEM had to ask Celestion to make different impedance versions rather than their usual standard 8Ω and 16Ω drivers. It's why you often see impedance badges on the back of earlier WEM cabinets reading 6Ω, 12Ω and 24Ω.'

It was the versatile G12 speaker, having been given a new lease of life thanks to the engineering nous of Les Ward, which was becoming the workhorse loudspeaker of the rapidly growing music scene. Its powerful output gave voice to big-name artists, first through the 'backline' amplifiers into which they plugged their instruments and then again as venues got bigger and the demand for 'front of house' sound systems increased exponentially. The time of the 'Power' speaker had come.

Introducing the DITTON 10

A high-fidelity full-range loudspeaker system (only 12¾" x 6¾" x 8¼") designed for situations where space is limited and yet the highest quality of reproduction is demanded.

Celestion Studio Series

8 HiFi Comes of Age

If the origins of high-fidelity music reproduction were rooted in the work that Celestion and others were experimenting with back in the 1920s, the term itself wasn't coined until three decades later. Its aim was to identify and describe equipment capable of rendering a highly accurate, low-noise and low-colouration reproduction of recorded audio, using exciting new methods like the new 33rpm vinyl long-playing record (first introduced in 1948) and the reel-to-reel tape in which Truvox was a specialist.

By the late 1950s HiFi was rapidly gaining popularity among audio hobbyists and enthusiasts up and down the country. In part, this was because of the availability of those new technologies that were able to render high-quality recordings, but perhaps more significantly it was due to the emergence of stereo recording, which had been winning round an initially reluctant public and was destined soon to overtake the more basic-sounding mono as the predominant recording format.

Technically, it could be argued that Celestion had already entered the domestic market at the end of the previous decade with Colaudio (a coaxial speaker featuring an HF device mounted to the centre of a woofer), which was intended for DIY enthusiasts who preferred to build their own equipment. This had proven popular enough to warrant an upgrade to a MkII version in 1963.

Rather than simply an updated version of the earlier coaxial speaker, however, Colaudio II had metamorphosed into a completed cabinet system. It featured an entirely new 12in bass speaker which, instead of featuring the conventional paper cone, had a diaphragm machined out of expanded polystyrene. The machining work was done in a dedicated and separate machine room in the Thames Ditton factory, where, according to eyewitness Bill Strong, it looked like it had 'continually been snowing'. The 12in bass unit was combined with the well-established HF1300 high-frequency pressure driver, and the whole cabinet handled 15W peak power.

Soon after, the TS100 and TS200 enclosed speakers were introduced under the Truvox brand (using Celestion-branded speakers) to complement the Series 100 and 200 tape recorders and a tuner amp.

Opposite: The Ditton 10 bookshelf speaker.

This is certainly a very musical loudspeaker and one for which I am sure Celestion will long be remembered

However, even with its rich pedigree in loudspeaker development for pre-war 'home entertainment systems'—the gramophones and radiograms on which the company had made its name—Celestion was aware that it had yet to introduce a product that could truly keep pace with the modern, 1960s idea of HiFi. With eyes firmly on that prize, the loudspeaker designers at Thames Ditton worked on creating the first of many notable Celestion HiFi loudspeaker designs.

The initial result was the Ditton 10, launched in 1964 and named for its birthplace and the modest power input it was capable of accepting (it was rated at 10W RMS). If not quite revolutionary, this speaker was certainly instrumental in defining what HiFi could be: no longer just a hobby for affluent devotees, but simply great-sounding music reproduction, now affordable for the masses.

The Ditton 10 had been designed to satisfy a demand from the market that had hitherto been neglected. Not only was it a genuinely high-efficiency bookshelf-sized product—essential, given that it had so few watts to play with—but it delivered a bass response that pundits had believed could only be achieved with a much larger and heavier enclosure. Beautifully finished in teak (part number T1172) or walnut (T1173) veneers, each cabinet contained a 5in woofer and a tweeter based on the HF1300 design.

The loudspeaker was stringently tested, and the results were reported by Geoffrey Horn in *The Gramophone* of June 1965, when the Ditton 10's price was listed as £18.18s. His assessment?

> This is certainly a very musical loudspeaker and one for which I am sure Celestion will long be remembered … I can have no hesitation in in putting forward the Ditton 10 as a loudspeaker which will particularly merit the gratitude of those who have little space, limited funds and high ideals.

If there was one particular jewel that emerged during that period, it was the HF1300. First developed by Les Ward for use by GEC (the General Electric Company), possibly in conjunction with GEC's Hugh Brittain, it was referred to as the 'Presence Unit' and used to extend the frequency output of GEC's metal cone loudspeaker. The design was likely adapted from Ward's earlier work on pressure drive units, which by this time had become his true speciality.

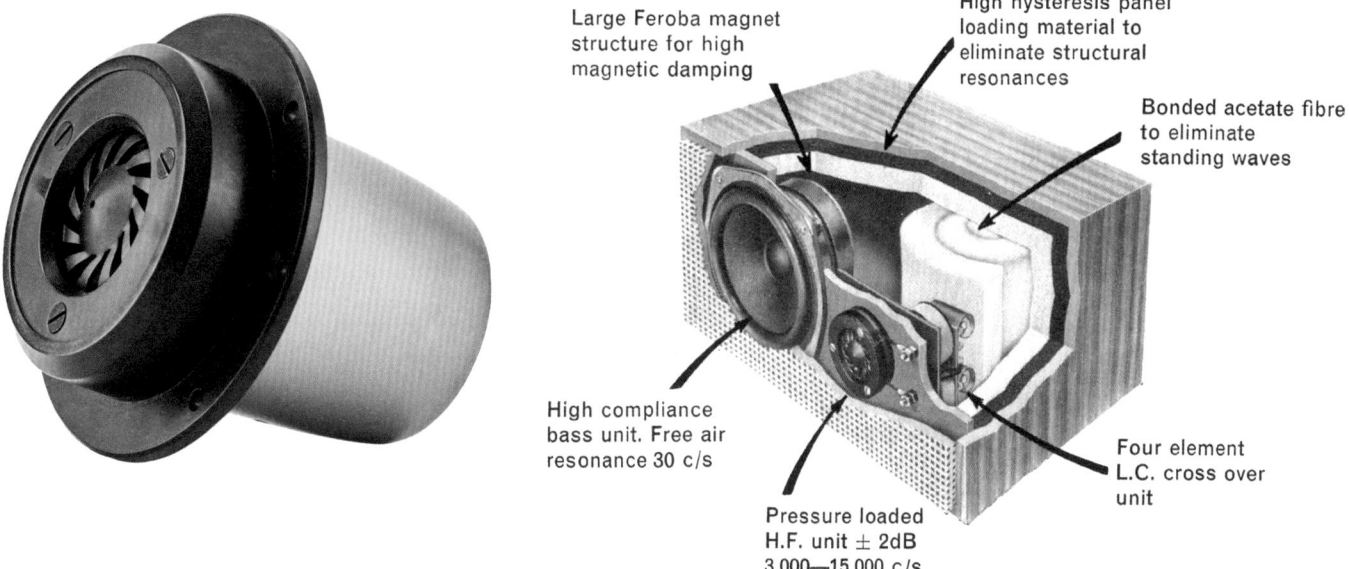

Left: The HF1300 pressure loaded tweeter.

Right: The inner workings of the Ditton 10.

HF1300 became a classic and enduring design, as important to the HiFi segment as the G12 was proving to be to the guitar speaker market that was booming simultaneously. Just like the G12, the HF1300 underwent further iterations and redeployments and was used by many other renowned HiFi speaker makers, including KEF, Spendor and Bowers & Wilkins. It was also used by the BBC, who fitted a version with a 34mm acoustically loaded rigid diaphragm into their studio monitors.

The HF1300 also became the template for several public address horn devices, including a police motorcycle handlebar speaker and AA and RAC (roadside breakdown recovery) motorcycle and sidecar outfits.

However successful the Ditton 10 was, there was still a sense that Celestion had arrived relatively late to the HiFi party. It was said they had lost several valuable years, largely due to a reluctance by the managing director at the time, Jim Tyrrell, to support the programme. It was only when ex-Plessey man Neil MacKinlay took over in 1965 that true momentum was created behind the move into HiFi, and staffing levels at Thames Ditton rose to around 400 in response to the growth.

Nevertheless, the Ditton 10 proved a winner and fired with this success, Celestion's design team continued the pursuit of still-greater sound quality. The result of this work was a completely unique idea, the Auxiliary Bass Radiator (ABR), the purpose of which was to extend the low end of the frequency response.

Left: Cutaway of the Ditton 15, showing the ABR.

Right: A fine pair of Ditton 15s.

An important technological development in the history of HiFi, the ABR was a passive unit—almost a loudspeaker with no motor—intended to work in conjunction with the bass speaker. It vibrated like a piston at low frequencies in sympathy with the main transducer to reinforce the output at the lowest end of the register.

It was an invention that made the world of HiFi sit up and take notice when the ensuing cabinet, the Ditton 15, reached the marketplace during 1966. With a larger 8in woofer, it duly confounded the experts with its considerable low-end response, disproportionate for such a small enclosure, while the attractive price—£28.11s.6d a pair—was certainly appealing to customers who were increasingly becoming more stereo savvy.

As a three-element system, the Ditton 15 easily met their aspirations. Bass response was substantially improved thanks to the ABR (extending the LF down to a purported 30Hz), as well as a newly designed 8in mid/bass unit. It also featured an updated and improved version of the HF1300 tweeter.

The Ditton 15 became the biggest-selling bookshelf loudspeaker of its time and remained in production for over a decade, during which period around 250,000 were made, while the product evolved into the Ditton 15XR with a slightly extended response range. Its success would also

> The Ditton 15 became the biggest selling bookshelf loudspeaker of its time and remained in production for over a decade, during which period around 250,000 were made

unlock export markets—for example, in France, where agent Universal Electronics, later purchased by Celestion to form the French subsidiary Celestion SARL, enjoyed outstanding success.

The Ditton 15 was in turn followed by the Ditton 25, which featured a wider bandwidth and higher power-handling capacity, along with increased output volume. Introduced in 1969, it was the first HiFi speaker to incorporate a specially developed HF2000 'supertweeter', which was intended to reproduce HF signals up to and beyond 20kHz, immediately making the output of the new model sound more detailed. Also developed by Les Ward, the HF2000 was capable of reproducing signals up to the astonishingly high frequency of 38kHz—well above the threshold of human hearing!

This of course was a marked improvement over the trusty HF1300 used on Dittons 10 and 15, which quickly rolled off signals above 15kHz. Instead, the Ditton 25 used a pair of HF1300s, wired in parallel, to reproduce the upper mid-range—from 2,500 to 9,000Hz—further adding to the detail and presence of the 25's sound.

At the other end of the scale was a new 12in bass reproducer, referred to as the UL12 and used in conjunction with an expanded 12in diameter ABR, making the whole system much larger than the Ditton 15 (whose bass unit and ABR were 8in) resulting in a low-end response that was truly impressive. As Swedish magazine *Hi-Fi Stereo* had it:

> The loudspeaker is a really good example of the auxiliary bass reflex principle, characterised by the distinguished presentation of the frequency operation, with noticeable good bass reproduction.

In five years, with just three key products, Celestion had carved itself a significant slice of the HiFi market. With this rapid success came more growing pains, and by the close of the decade Celestion had finally, unquestionably, outgrown its factory in Thames Ditton and it was time to leave. Now established as a marque in its own right, however, the name Ditton was set to live on.

9 Celestion Industries, plc

The 1960s was arguably the decade when Celestion really began to hit its stride. The Truvox brand began to take more of a back seat as a portfolio of industrial and public address products were marketed under the Celestion PA sub-brand. The 1967 *Loudspeakers for Public Address* catalogue proudly stated that its featured loudspeakers 'are in service with British and foreign governments, leading international public address contractors, police, ambulance, fire services and airport authorities'. It included re-entrant horn loudspeakers familiar from the Truvox range, alongside Les Ward-designed high-efficiency MF/HF pressure drive units (an early example of the modern-day compression horn driver).

In a nod towards the emerging sound reinforcement market, also included in the catalogue were complete column loudspeaker systems. Featuring either seven 8in x 6in speakers or seven 10in speakers, they were sold as 'suitable for use with both speech and music signals', which 'provide extremely good quality reproduction locations where ambient noise levels are high'.

Space for production was once again becoming a problem at Ferry Works because of the success of the G12 guitar speakers—with customers clamouring to receive their order as soon as possible—alongside the mounting demand for Ditton HiFi speakers. This popularity resulted in further strain being placed on a factory already at almost full capacity trying to satisfy the demand for the sizeable commercial and industrial speaker portfolio.

Ultimately, as expansion at the Thames Ditton site was restricted by the river on one side and a major road on the other, a different location was deemed the only viable solution. Problems had also been encountered with various docks in the London area, so eyes were cast further afield: to East Anglia, a region 100 miles or so to the north and east of London. The area was surveyed and after a plan to settle in Bury St Edmunds, Suffolk, originally announced to staff in June 1968, was summarily abandoned, an alternative site was acquired on Foxhall Road, on the eastern side of nearby Ipswich.

Opposite: A full catalogue of public address speakers.

A regional hub for manufacturing, Ipswich had a ready-made supply of engineering and assembly workers available to augment those who made the move from Ferry Works. Direct road and rail links meant that transportation to and from London and the South East, as well as central England, was straightforward, while sea freight could be handled from the nearby port of Felixstowe.

Thus began the gradual transition to a new facility, fittingly to be named 'Ditton Works', perpetuating the legacy built from two decades of manufacturing by the River Thames. Bob Smith, who joined the company from Goodmans just as it was setting out the plan to move to Suffolk, remembered, 'The Ditton 10 and 15 were already heavily in production and the Ditton 25 was about to come out. And when production on that started to increase, they had to take over another room—the Listening Room—where they did the cabinet assembly.' Such was the lack of space in Thames Ditton.

Smith continued: 'Most of the Celestion measurement equipment was antiquated back then, with GPO Induction and Capacitor system dating back to the 40s or early 50s used for measuring voice coils and components. It wasn't until the move to Ipswich that the Brüel & Kjær system was fully brought in to measure the acoustics side, response and distortion.'

In the wider world of Truvox, Daniel Prenn clearly had corporate consolidation on his mind during the latter part of the 1960s. When the lease on the Truvox factory in Neasden, North-West London, expired in 1967, Prenn merged one Truvox subsidiary with another of his companies, Thermionic Products. Production was moved to Hythe, near Southampton on England's south coast, which was also used as the base for the Truvox Floorcraft company, created at the same time to take sole ownership of the popular Truvox floor-cleaning products.

That same year, another of Prenn's companies, Control and Communications Group, acquired Truvox tape recorders and the associated business, again from Truvox (Neasden). In 1969 that company was in turn acquired by Racal Electronics, which at its height was the third largest British electronics firm, resulting in Prenn becoming a significant Racal shareholder.

It can be no coincidence that 1969 was also to prove a pivotal year at a corporate level in Celestion, with the completion of a reverse takeover of the publicly listed Weingarten Brothers clothing company—another of Prenn's business interests, in which he owned a 20% stake—by the now much slimmer and still privately owned Truvox Engineering, the nominal owner of Rola Celestion. The newly created entity was known as Celestion Industries, plc., and it was listed on the London Stock Exchange.

This secured new funding for the company right at the time when the move to the new premises was in progress and, rather oddly, added several clothing lines to Celestion's portfolio of products, including a range of ladies' 'intimate apparel'.

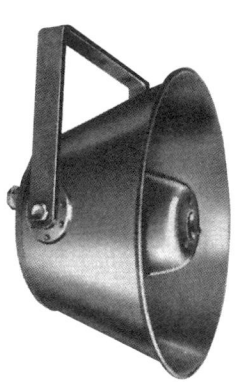   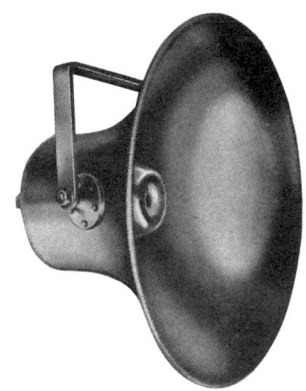

The public address range included robust re-entrant loudspeakers."

It was clear that, to the newly formed corporation, the clothing part of the business was perhaps even more important than the loudspeaker part. In May 1970 the *Financial Times* reported that the first expansion move created by this merger was the incorporation of another garment company, Bristol-based Chappell Allen. In 1977 Celestion also went on to buy Derbyshire-based Wood Bastow; both new companies produced ladies' underwear.

At Foxhall Road new buildings had been constructed and existing ones modified, and the nucleus of a workforce had been recruited using ex-Thames Ditton supervisory staff to train new employees from the local area. Production had begun in Ipswich in late December 1968, and that had marked the beginning of a phased transition between the two factories.

First to transfer to the new Ditton Works was the 'Power' line (so-called due to these speakers' ability to withstand greater power input): permanent-magnet speakers that had evolved from Les Ward's toughened G12 that had been used to such great effect by guitar amplifier and PA companies, now expanded into a full range of products. This was later followed by domestic speaker lines—the Ditton HiFi cabinets destined for people's homes—and finally the industrial and commercial speakers.

With this new injection of cash and a new premises, confidence was high, and success was in the air. So much so that an extension to the Foxhall Road building had immediately been required, and this was duly completed in the spring of 1970.

A new company and a new location befitted the start of a new decade. So impressive were the turnover and pre-tax profits in 1970—well above forecast—that the board's confidence was reflected by the announcement of an early dividend for shareholders, with advertisements being taken in several publications including Britain's newspaper of record, *The Times*, on 30 September, announcing Celestion Industries' success to the world.

3

Power to the People

Live Aid: among many other things, a significant moment in live music history.

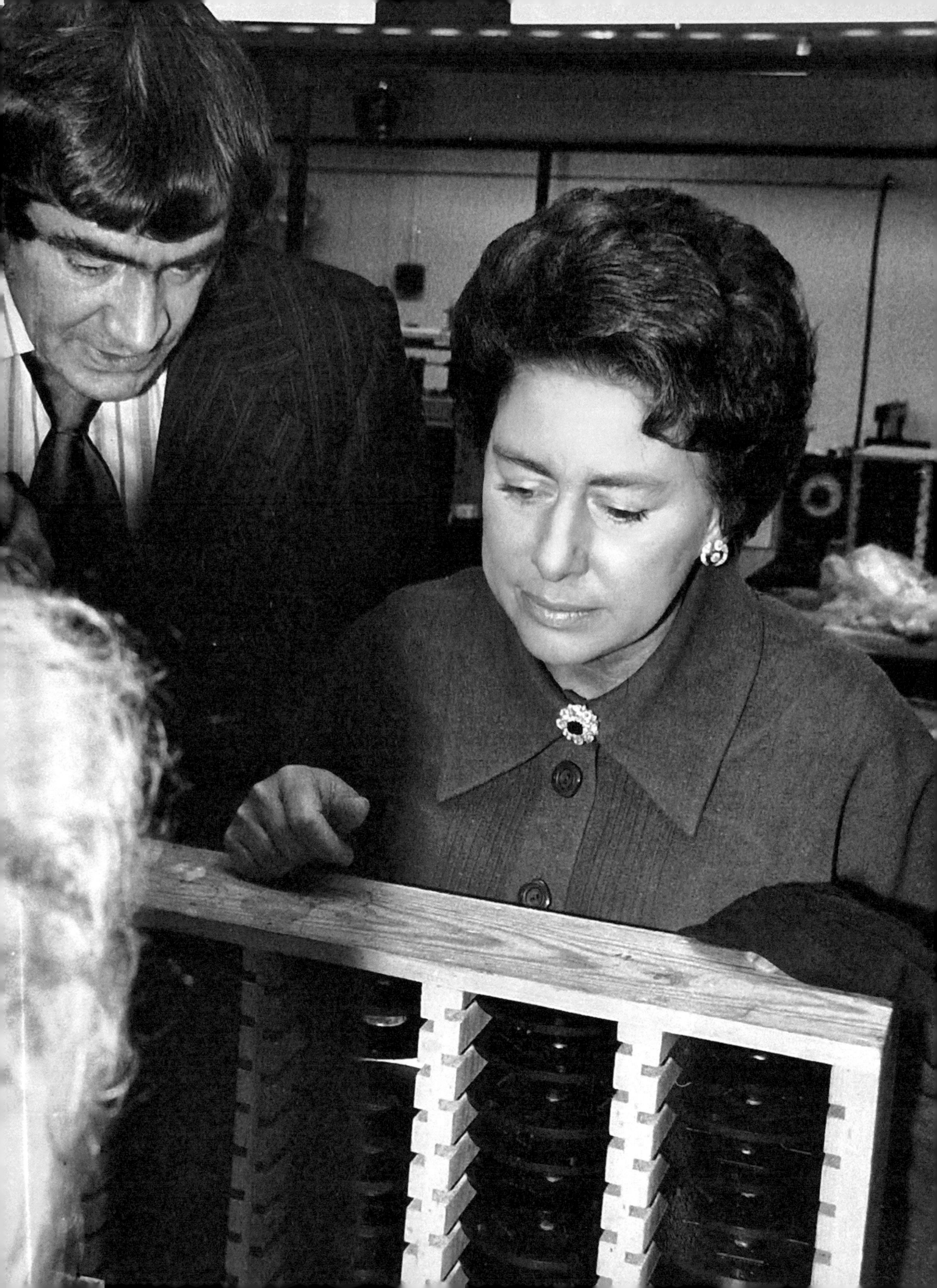

10 Two Decades of Turbulence

During the early days following the transfer between old and new factories, product supply and distribution raised challenges. It was often necessary to dispatch a vehicle early in the morning from Ipswich to make the 90-mile run through London to Thames Ditton, carrying the previous day's production run of drive units for HiFi speakers, ready to be loaded into cabinets. The vehicle then loaded up and returned by the same route, bringing back essential components for the following day's production, enabling the supervisor, charge hand and production line to meet their schedule.

As the weeks passed, the interior of the Ipswich works began to house ever more sophisticated production machinery, and more staff were engaged, first to be trained, and then to operate the lines, which were now able to relieve some of the supply problems that had been experienced at Thames Ditton. 'They did it very well although it was chaos for a while,' recalled Bob Smith.

Logistically, where to house which production—let alone the respective research and development (R&D) and admin support—must have created a nightmare in those early days for the Rola Celestion management, which was now headed by Managing Director Neil MacKinlay and included Production Manager Jeff Rooker, who was heavily involved in the initial factory move. Rooker, who departed soon after the move was completed, went on to become a UK Member of Parliament and subsequently sat in the House of Lords.

The relocation brought other benefits, including better measuring systems. 'Previously we would just sweep a product over an oscillator and apply a lot of guesswork, but for the Ipswich factory we purchased some Brüel & Kjær analysers and measurement mics for the production line, with a CRT [Cathode–Ray Tube] monitor,' Bob Smith added. It was all top-of-the-range equipment, meant for high-quality manufacturing.

Another of the brains behind this testing set-up had been experienced engineer Laurie Fincham. He had been instrumental in the development of the early Ditton speakers and had been responsible for bringing in the Brüel & Kjær equipment.

Opposite: HRH Princess Margaret inspects the Ditton tweeter production line.

Dave Robinson operates the Brüel & Kjær measuring equipment.

Fincham also introduced Smith to Charlie Watkins of WEM and Selmer's John Hosey. 'I still went to a lot of live shows and consequently became more involved in the PA side,' said Smith, recalling that he would undertake third-octave band measurements on the powerful musical instrument and PA system drive units for Technical Director Les Ward. Although still mainly involved in what was increasingly being referred to as the 'professional audio' business, at almost 60 years old, Ward was less inclined to be directly involved with live music.

Smith's impression of Ward was of a military type: 'He was very upright, always smartly dressed, and with a handlebar moustache; he could have been a squadron leader. Les was an old-time engineer who had learnt from experience rather than having gone to university. He had a background working with musical instrument speakers and was known everywhere for having developed the Vox Blue.'

Ward's main area of expertise, however, was with pressure drive units for mid-range and high-frequency reproduction. Early developments had included flameproof speakers for mass notification in hazardous environments, which remained in production through the 1970s. He'd been responsible for the development of the HF1300 and HF2000 treble units used in many HiFi cabinets. He had also developed the MH500 (which had generated a patent in 1973)

Daniel Prenn presents Les Ward with a Morgan sports car on the occasion of his retirement.

and MH1000 mid/high pressure drivers, which had been all but ubiquitous in the sound systems of the early 70s; in fact, Ward had also designed a horn for use with the MH drivers that had been sold to Marshall for use in their PA cabinet range.

Smith recalled Les Ward retiring from the company gradually, initially moving to a three-day week, before taking on consultancy work at the company. When he did finally call it a day, in 1979, he was gifted a bright red Morgan sports car by Daniel Prenn, apparently to 'replace his Volvo'.

The incorporation of the company as Celestion Industries and its resultant appearance on the London stock market necessitated regular reporting of financial information, and in the early 1970s company results were impressive. After beating its forecast by 54% in its first fiscal year, in 1970–1971 Celestion lifted pre-tax profits by a further 71%—and this even though Foxhall Road was still running some 25% below full production capacity. Pre-tax profits were higher than anticipated, with the impetus said to be still coming from the industrial public address business, despite the runaway success of guitar speakers and Ditton HiFi.

Profits continued to rise within the Group, the £275,000 declared in 1971–1972 easily eclipsed by the £480,000 pre-tax profits showing on the following year's balance sheets. But it was the

The Foxhall Road site.
Inset: Recruitment ad for the Hadleigh Road production lines.

'specials' sector of HiFi that was beginning to pique Prenn's interest, as he promised shareholders that the company would be diversifying away from 'bread-and-butter' items with the increasing demand for more sophisticated stereo separates. The truth was that the production of domestic speakers had reached a peak about three years after the move to Ipswich, and there was a perceptible shift in the market, with a noticeable decline in entry-level product (where Celestion dominated), which was offset by the clamour for much more high-fidelity components.

After a couple of years of rapid expansion, Foxhall Road was bursting at the seams, so the search began for a second property nearby. Dedicated to HiFi, this was to be a thoroughly modern assembly plant, and a location was found on the Hadleigh Road Trading Estate. This new factory became operational early in 1973 and was expected to meet current demand and then expand to maintain the northward direction of growth. Situated on the other side of Ipswich from Foxhall Road, Hadleigh Road was to be 'capable of making a growing contribution to turnover and profitability', Prenn told shareholders in July that year.

Meanwhile, the transfer from South-West London was still far from complete. In a memo to staff dated July 1973, Managing Director MacKinlay revealed the necessity of continuing to manufacture at Thames Ditton, owing to a sustained increase in demand. He announced that

Staffing levels at Celestion reached a peak in January 1974, with a total of 907 employees across the three sites

the factory would run for an unlimited period on production of public address speakers, cones and suspensions, with a further cone tank being prepared for installation. The plating department and coil winding would likewise remain.

In 1974 turnover expanded from £5.42m to £6.23m, while pre-tax profits to 31 March increased again, to £532,560. This was the case even though the clothing side had been badly affected by the government-imposed three-day working week. At this point, famous British department store Marks & Spencer was now responsible for 80% of demand for Celestion clothing production. It was left to the loudspeaker division to protect the balance sheet, and, thanks to its export business, this jumped by 60% to almost £1.2m in sales.

In the country at large, the three-day week created socio-economic and political mayhem. It had been implemented as a consequence of the oil embargo imposed at the end of the previous year by OPEC (the cartel of oil-producing nations) but in truth the seeds were sown as far back as 1972, when Heath's Conservative Government attempted to place a cap on public sector wages to curb inflation. This ultimately led to industrial action by coal miners, which resulted in a fuel shortage—hence, in 1974, with the nation's electricity supply rationed, Prime Minister Heath had no option but to mandate a shorter working week.

Staffing levels at Celestion reached a peak in January 1974, with a total of 907 employees across the three sites, some of these working night shifts, and HiFi sales showing no signs of abating. Distribution was now wholly carried out from Ipswich by a new fleet of company vehicles. A new marketing policy saw the company opening up a great many more smaller distribution points instead of the previous few selected wholesalers. However, this boom was short-lived, and by August 1974 there was enough of a reduction in demand that capacity at Ipswich finally became available, enabling the remaining production to be transferred from Thames Ditton.

The Foxhall Road site was already starting to look tired even before the transition of production from Ferry Works was complete. One employee who remembers it well is Dee Potter, who arrived at the factory on a Monday morning in April 1974 aged just 16, having left school the previous Friday.

Dee started on the assembly line making tiny diaphragms for customer Derby Autos; however, the factory didn't quite make the impression she'd been hoping for: 'The day I left school I told the headmaster I was going to Rola Celestion, and he said, "I've heard life is quite boring there."

TWO DECADES OF TURBULENCE

But I wanted to leave school early and joined the company despite living three miles away. My mum and several of my family worked there at the time, and my mother got me the job to work on PA products. ... The place was huge and very dirty. When it rained it came in from both the top and bottom. It certainly wasn't a glamorous place to work!' But Dee didn't seem to be phased by this, and just as Celestion celebrates its centenary in 2024, so Dee Potter will reach her own landmark anniversary, marking 50 years' service.

Celestion's involvement in the clothing industry produced some unexpected fringe benefits, Dee remembers. 'On a Friday they used to open up a shop on the side of the factory floor and you could buy ladies clothing and underwear at a discounted price.'

Joining the manufacturing staff the same year as Ms Potter was another long-serving employee, who eventually clocked up 27 years with the company. When Brian Brame arrived in 1974, his focus was on the Hadleigh Road premises. 'I helped redesign the interior with manufacturing director, Colin Aldridge,' he remembered.

As a junior engineer, Brame also had clear memories of Foxhall Road: 'When I first arrived, we were still operating on one assembly line. We were producing some flameproof products and we had the Schutte lathes which were making magnets for tiny elliptical speakers for TVs. The last thing TV manufacturers ever thought about was the speakers! We'd make thousands of them in a day, people working on either side of the production line, performing the same operation.' He remembers the Hadleigh Road factory being completely HiFi-focused: 'We also made crossover networks and had a tweeter assembly room, but the HiFi bass units were still produced over at Foxhall Road.'

1974 was also the year that saw the Ditton 66 arrive in the marketplace. Ultimately, there would be three production lines dedicated to HiFi alone, with night shifts operating until 10pm. While speaker manufacturing for the TV and radio sector was coming to the end of its lifespan, products like the Ditton range were hitting their peak.

The original Hadleigh Road set-up had been based on a single production line with an anechoic chamber, but success brought a major investment and led to a complete upgrade. As part of this modernisation, there were instead three cabinet lines with their own dedicated anechoic chambers, each patrolled by Bob Smith and Dave Robinson on the technical side with Brian Brame overseeing construction: all complete with Brüel & Kjær measurement equipment, including oscillators and display screens. Above, a mezzanine was added to store materials.

The refurbishment of the site was completed during the summer of 1976 and on 29 October it was officially re-opened by Her Royal Highness, Princess Margaret (younger sister of Queen Elizabeth II).

HRH Princess Margaret officially opens the newly refurbished Hadleigh Road factory, October 1976.

This upgrade in tooling and infrastructure was tacitly referred to during the results announcement for 1976–1977. Group profit had continued to rise consistently through the mid-1970s—and so it was with great pride that Chairman Daniel Prenn was able to announce that performance up to March 1977 showed that turnover had risen by 41.9% to £12.3m and pre-tax profits passed £1m for the first time (at £1.09m), a 66.4% hike. The trading improvement had come from the audio side (rather than clothing), said the report. Aided by a factory re-equipping (the Hadleigh Road refurbishment), there was a rise in loudspeaker sales—and it was a considerable one, heading up from £4.4m to £6.2m.

Despite this ongoing success, Prenn and Managing Director John Church, who had succeeded Neil MacKinlay, could see that export markets were undernourished, with opportunities not exploited to their maximum. Consequently, two subsidiaries were set up in Europe at the end of 1976, with a plan for a third subsidiary in the United States to follow. Of the future, Prenn stated that the overseas subsidiaries formed in France, West Germany and the United States 'were expected to make a growing contribution to profits'. The increase in geographical outreach led to the adoption of the name Celestion International Ltd as a corporate identity for the loudspeaker division of the company and all its major overseas subsidiaries.

If HiFi and guitar chassis speakers represented the more glamorous side of the business, sales in the industrial and commercial public address sector remained highly significant. For example, the Royal National Lifeboat Institution specified products from Celestion's public address division for their durability in arduous situations. They were also extensively used by the railways, merchant shipping, Armed Forces, oil refineries, coal mines and the emergency services—particularly police motorcycles—as well as for factory installations, telephone systems and sporting events. Special explosion-proof units were available for operating in hazardous conditions, while glass-fibre speakers would cater for corrosive and damp applications. So it was that all the announcements at the Dounreay nuclear power establishment in Scotland came to be made through Celestion PA.

Celestion public address found applications varying from airline cabin communications using specialised cone loudspeakers to shipboard deck communications, as well as natural gas drilling rigs, fire engines and factory installation using weatherproof re-entrant loudspeakers. The installations on the offshore gas production platform incorporated 'talk back', which enabled the loudspeaker to be used as a microphone for two-way communication. This dominance of the commercial public address market only ended in the 1980s with the arrival of a leading Japanese brand, which then proceeded to gain control of the market.

Reflecting on the late 70s period in an early 1982 interview with the *Financial Times* newspaper, Colin Aldridge, who by then had taken over from Church as managing director, disclosed that he felt that Celestion had begun to lose ground in the race for technological development during the HiFi boom. He felt that the company 'was faced in 1978 with an ageing product range and a shrinking market'. In response, Celestion 'had decided on a product-led recovery with investment in R&D but ran into a crippling recession from the end of 1979 and their reaction was to close the Hadleigh Road factory'.

The beginning of the 1980s was to deal Celestion a bad hand, as it did to the whole of British industry. Prices had begun to spiral upwards as inflation increased by more than 20%, while an energy crisis akin to the mid-70s again drove up oil prices. The combination of these pressures, along with high interest rates, compounded by the strength of sterling, produced a perfect storm. Celestion's masterplan, hatched several years earlier to dominate export markets, was now stalling. Group profits were hit, in particular the loudspeaker division. By March 1980 the balance sheet had turned from black to red, a profit of £37,000 now showing a loss of £525,000. Market conditions were described as 'dismal'.

Left: Dounreay Nuclear Power Establishment and RNLI Lifeboat 'Tony Vandervell'; both with Celestion public address speakers.

Right: Public address product catalogue, 1984.

This situation was to get worse, and by the time the company declared its next half-yearly results in September, it was deeper in the red, having surged to a loss of £1.16m under the weight of these catastrophic economic factors, and their impact on consumer demand. The government's tight fiscal policy, designed to rapidly control inflation, continued to bite throughout 1981.

Given the unexpected change in market conditions, Celestion had to take a hard look at its approach to HiFi, as Colin Aldridge had pointed out in that *Financial Times* interview. 'We had to decide between the all-singing, all-dancing speaker, at whatever cost, and developing a really HiFi quality product, which many people could afford. And it would serve as our flagship,' said Aldridge. Enter the game-changing SL6, retailing at just £250 a pair, to rave reviews from the HiFi press. It would prove to be a bellwether in the HiFi world, and one which owed its success to the company's forward-thinking investment in laser interferometry.

As far back as 1978, the company had begun perfecting a laser instrument which, when linked to a computer, would measure the fine vibrations of the speaker, and by the beginning of the 80s

Technical Director Graham Bank, along with Gordon Hathaway, an engineer, had developed the system that allowed visual examination of diaphragm break-up modes in a unique manner.

With great ingenuity it was Bank and Hathaway who adapted a BBC microcomputer to control the laser scanning equipment, enabling a system whereby HiFi, public address and music products could be measured in detail, without the need for a more powerful multi-room computer set-up, which at the time was prohibitively expensive. Soon afterwards, Celestion graduated to more sophisticated computers as the technology became more accessible and, with the arrival of Julian Wright in 1984, further sophisticated design programmes and measurement systems were developed.

Although industrial public address products had all but tailed off by now, the sound reinforcement business remained strong, and by the mid-1980s inroads had been made into the development of a new kind of finished sound reinforcement system. Celestion had tinkered around the edges of this market for some time, but the highly original SR system represented a serious investment and the company's first attempt at staking a real claim in this sector. Making use of the now-indispensable laser interferometry equipment in the design of the SR's forward-thinking moulded cabinets, and in the unique coaxial drivers that were used inside those cabinets, SR system development combined the expertise of the HiFi and 'Power' speaker camps within the company, and with that co-operation came a new channel of business to explore.

In 1984, after three years with a negative balance sheet, Celestion had swung from a £610,000 loss back to £466,000 profit. Success was fleeting, however, and by 1987 profits had slid back again. That same year, in a bid to halt any further slide, Gordon Provan was headhunted as the company's new managing director with the task of taking the company and the three overseas subsidiaries and forming them into a stronger and more cohesive group. Straight away he changed the management of the subsidiaries in North America and West Germany and relocated them so that they started trading more profitably.

The fundamentals were already there, with a strong product offering and a global reach. However, Provan was determined to make the company more sales- and marketing-focused than some of his predecessors, bridging the gap between the engineering know-how that was clearly present within the organisation and the demonstrably increased sales success that he believed was readily achievable.

Celestion had survived two decades of turbulence as a market-listed company: battered by political and financial headwinds, tethered to a larger subsidiary whose performance in the fickle clothing industry had been variable. Happily, the company had come out on the other side in good health, and, with a new leader and a new focus on HiFi, music-making equipment and sound reinforcement systems, had a strong vision of the future.

The Prenn Dynasty

From left to right: Alexis, Oliver, Daniel and John Prenn, *c.* 1976.

In 1986, Richard Klein joined Celestion as a technician and worked his way up first to junior engineer and then to OEM development manager, before later becoming production manager and clocking up more than 30 years' service along the way. At his initial job interview he remembers seeing a locked door in the development lab, which he later discovered was Daniel Prenn's private bathroom, while a young man beavering away nearby on the shop floor turned out to be another member of the Prenn dynasty, Daniel's grandson Alexis (who later went on to become Celestion's USA sales director for a short time). 'I wondered what I had walked into,' Klein exclaimed.

Celestion Industries at this point was very much a family business. Daniel Prenn's oldest son Oliver (father of Alexis) had been born to his first wife Charlotte in September 1937. Some 20 years later, after separating from Charlotte, he had a second son, John, with Maxwell A. Taylor, with whom he lived until her untimely death at the age of 42. Both sons had been directors of the Celestion Industries, with Oliver in particular taking an active role in the loudspeaker business during the 1980s.

By October 1988, Daniel Prenn was clearly ready to call time on his distinguished 40-year association with the company—perhaps to spend more time with his beloved racehorses—making way for Oliver as chairman. He passed away three years later at the age of 86.

11 The Golden Decades of HiFi

HiFi production really took off after the runaway success of the Ditton 15, following the slow migration out of Thames Ditton to Ipswich. One of Bob Smith's first tasks after the move had been to upgrade the Ditton 10 to a MkII version, replacing the venerable HF1300 with an updated version, and, owing to the arrival of transistor amps, changing the speaker's typical impedance from 15Ω to 4–8Ω.

Along with fellow engineer Dave Robinson, Smith was also closely involved in the development of subsequent Ditton models, most notably the Ditton 44 and 66, both of which garnered highly favourable reviews after their launch.

The Ditton 44 Monitor had been introduced in 1973 and featured the Les Ward-designed HF2000 tweeter that had originally debuted in 1969's Ditton 25, plus a 5in mid-range cone speaker (which was the woofer from the original Ditton 10, adapted by being sprayed with a surface coating), and the tried and tested 12in UL12 bass unit, all built into a large cabinet for additional heft.

According to the sales leaflet, it was 'designed for the discriminating listener ... with a wide, smooth frequency response. Exceptional transient performance, superbly controlled bass, accurate mid-range and smooth extended highs.' It had been reviewed by John Gilbert for *The Gramophone,* who enjoyed the stereo imaging of the speakers, adding that 'one noticeable feature was the forward tone, missing in some enclosures. One felt that the performers were in front of the speakers instead of being heard some distance behind.'

Following in 1974, the Ditton 66 was described as: 'A new loudspeaker of advanced design suitable for studio use and for home installations of the highest quality.' In common with the Ditton 44, it also featured an HF2000 tweeter and UL12 bass unit. Where it differed was in the mid-range, which instead was delivered by the Les Ward-patented MD500 (a 2in soft dome

Opposite: Julian Wright demonstrates the use of laser interferometry with HiFi cabinets.

Left: Rick Wakeman and the Ditton 551.

Right: Advertising for the Ditton Series.

pressure driver of 'advanced design'). The Ditton 66 also featured an Auxiliary Bass Radiator for additional bass extension. With all this equipment, tied together with a precision crossover, it's no wonder this behemoth of a loudspeaker had a cabinet that measured over a metre (40in) tall.

The sound was phenomenal too. According to *HiFi Answers*, 'We can say very little to criticise the speakers. They are large, expensive and very accurate!'

Along with these two flagship products, by now the range featured the well-established Ditton 15 and 25, the upgraded Ditton 10 Mk II, the bookshelf format Ditton 120 (essentially a Ditton 15 with a higher power capability) introduced around 1971, plus the compact, entry-level 'Hadleigh' and 'County' models. Celestion now had cabinets that covered a comprehensive range of performance and price points, or, as the brochure would have it, 'true high fidelity for the impecunious enthusiast or the affluent afficionado'.

One curious innovation that was also available as part of the range was 'Telefi'. Connecting to a television wirelessly via inductive coupling, it enabled viewers to listen to the TV's audio signal through their HiFi speakers, ensuring 'crisp, full-range, distortion-free reproduction of music and speech providing an improvement over ordinary TV sound which will amaze you'.

The UL6 won a 'Speaker of the Year' award from the *HI-FI Choice* Awards when it was introduced

These models were followed later in the decade by more Ditton variants, including the 121, 332, 442, 551 and 662. So well thought of was the range, that subsequent permutations of Ditton speakers continued throughout the 1980s (including 'Series II' versions of the 44 and 66), with the last of the Ditton line launched as late as 1990.

Following the success of the first generation of Dittons, Celestion introduced the UL range of 'studio quality bookshelf speakers' in the mid-1970s, for which Dave Robinson developed a mid/bass unit using a 'bextrene' cone instead of a paper cone—a composite thermoformed plastic that was already known for having been used in BBC monitor speakers. The bextrene woofers were said to produce a particularly neutral, uncoloured output, hence the name given to the range: Ultra Linear. The UL6 won a 'Speaker of the Year' award from the *HI-FI Choice* Awards when it was introduced in 1976. It also garnered accolades in other parts of the world, including Japan, a country where the company was determined to make inroads.

The year 1976 also saw the arrival of Dr Graham Bank, from Rank Radio International, as technical manager. Ultimately, Bank would take the technical director's reins from Les Ward following the latter's retirement, and by 1981 Bank's role had expanded massively, with responsibility for all product design and development. That included HiFi and the more business-to-business-focused professional audio products, which encompassed the musical instrument, sound reinforcement and public address lines.

Ed Form arrived later that same year and took on much of the professional audio mission: Bank's in-tray was overflowing, and it was a sensible division of labour that left Bank better able to concentrate on HiFi, which had been his abiding passion. Early in his tenure, his research had led him to developing, together with engineer Gordon Hathaway, Celestion's laser interferometry system, which allowed visual examination of diaphragm break-up modes in a unique manner. It was an innovation that was to have significant impact on subsequent generations of product.

A paper published in the *Journal of the Audio Engineering Society* (May 1981) described the process:

> When the beam from a laser vibration interferometer is optically raster scanned over a vibrating surface, a phase-sensitive detector provides velocity information at any phase of the motion. These data are digitally processed, and a hard copy print gives a three-dimensional isometric view of the complete vibrating surface of the test object, frozen in time.

The ground-breaking SL6.

During a time when Managing Director Aldridge believed HiFi innovation had begun to lose ground, a timely injection of talent and ingenuity was helping to reinvigorate Celestion R&D. And then came the SL6.

The SL6 and the Metal Dome Tweeter
The turning point on the journey towards developing the SL6, believed Bob Smith, was in 1980, when engineer Boaz Elieli joined the company, immediately playing a major role in the design of what became recognised as a ground-breaking loudspeaker. 'This was the point where people took Celestion HiFi seriously once again,' agreed Ed Form, 'because the SL6 was astonishing.' It was a product that would set the company off on a completely new path.

The SL6 was launched at the Harrogate Hi-Fi Show of 1981 to great acclaim from public and media alike. 'The engineers from Celestion used some pretty advanced techniques, such as optical interferometry to get the best from a relatively small 2-way speaker,' Trevor Attewell wrote in *HiFi News*. He continued:

> It is no exaggeration to state that the SL6 is one of the most interesting moving-coil speakers to come my way for a long time, and that it embodies significant advances in driver design. Celestion has traditionally been associated with the mass end of the

Close up of the copper dome tweeter from an SL600Si.

market, its reputation justifiably built on product consistency and value-for-money rather than on innovation.

Meanwhile, John Browning (*Vintage Sonics*) went further:

If there was ever a HiFi product that drew an invisible line under the 1970s, as if to declare, 'this chapter is now finished,' this was it. It's difficult to put into words how modern it appeared when it was introduced in September 1981; its design was so stunning that it became shorthand for all that modern loudspeakers were. Nothing would ever look or sound the same because it couldn't.

Elieli had maintained that soft domes, like phenolic or coated fabric, were incapable of truly accurate performance: to stop the dome from producing unwanted distortion it had to be rigid. He further reasoned that the dome surround was equally prone to inaccuracies, and concluded that the rigid dome should be bigger and the surround smaller for best results. His practical application of this thinking used a 32mm electro-formed copper dome and a very narrow surround – the same overall radiating area as typical 25mm soft domes, but with a large proportion of their 'wobbling-jelly' distortions suppressed.

THE GOLDEN DECADES OF HIFI

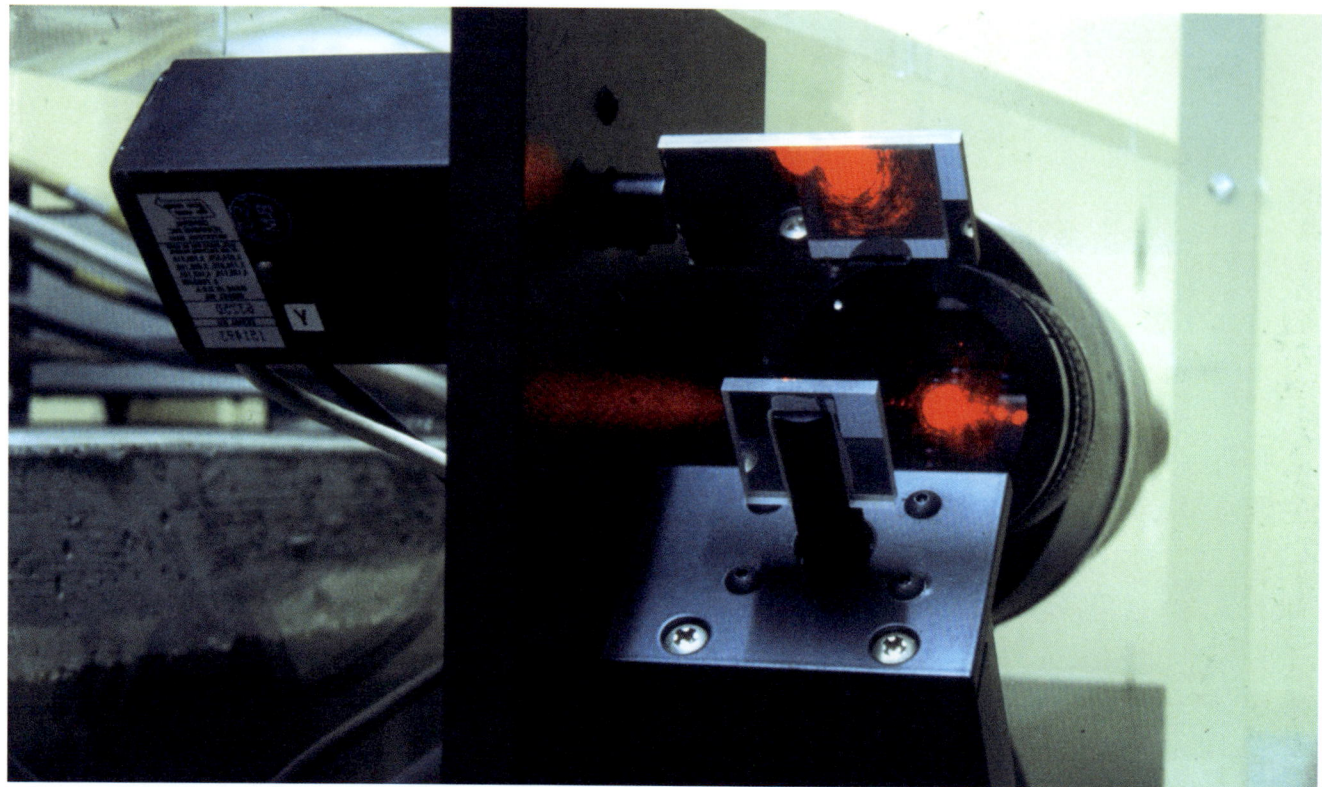

Close up of the scanning laser used for interferometry measurement.

It had been a radical development to base those first tweeters around metal domes. The company had originally been looking at silver but decided it was too expensive, so opted for copper as a viable alternative. The copper dome design was granted a patent in 1982 for Elieli following the launch of the SL6, just one of many transducer patents in which he was involved.

The SL6 also featured an innovative mid/bass unit, likewise developed by Elieli, a 6.5in cone in sculpted PVC, a new cone material for the time. Its shape, with an integrated, concave dust cap, unique suspension and precision-located voice coil termination, were all factors that actively contributed to lower distortion.

So had the SL6 truly put a lid on the 1970s with its advanced technology? History suggests that it did. And yet it wasn't perfect, with many believing that the original design had suffered from a lack of lower-mid-range transparency and a depressed HF response, or, as one commentator put it, 'it was too warm, without enough HF'.

These perceived wrinkles in the design were ironed out in the upgraded SL6S, which boasted a new metal dome HF unit, this time formed from a strip of aluminium. An improved crossover had been designed by Bob Smith for a later revision, the SL6Si, using more expensive polypropylene components and featuring a 12dB/octave low-pass slope for the woofer with an 18dB/octave high-pass function for the tweeter.

The payoff was an award-winning series that would continue to dominate the HiFi landscape between 1982 and 1988

The SL6 and subsequent additions to the range (SL6S, SL600, SL700, SL6Si, SL600Si and SL12Si) had all featured advanced and innovative technologies that wouldn't have been so easily in reach without the laser interferometry system, which was used to scan the cabinet and drive unit surfaces to produce a moving, three-dimensional image of their vibration pattern. The payoff was an award-winning series that would continue to dominate the HiFi landscape between 1982 and 1988. The SL600, in particular, set new standards at the audiophile end of the market. It sold exceptionally well in Japan and won numerous awards around the world.

In the SL range's development, Bank and Hathaway had decided that a different approach to cabinet design was required. What followed in 1983, in their quest to perfect the SL design, was a hugely expensive investment in a material from the aviation industry called Aerolam™, which was subsequently used to build the SL600 cabinet with the goal of removing much of the unwanted colouration thought to be inherent in conventional wood cabinets.

Aerolam was essentially an aluminium honeycomb sandwiched between thin aluminium walls and normally used in aircraft. It weighed about 25% of an equivalent sheet of aluminium but had almost the same stiffness. Ed Form recalled, 'I was experimenting with a relative of Aerolam to make small loudspeaker pistons and Graham, looking for a way to defeat the poisonous resonant behaviour of wooden cabinets, asked if it was possible to make cabinets from it. It wasn't possible with the material I was using but I persuaded [Swiss Chemical Company] Ciba-Geigy at Duxford to let me have some offcuts of Aerolam, and Les Surman and Bill Strong made them into a cabinet.'

As one review described it,

> The Celestion SL600 speakers provide outstanding imaging capabilities with an incredibly smooth and detailed midrange and a fast, low fundamental which is enough for all but the biggest orchestral works. The midrange is absolutely superb and the treble very controlled. The bass extension isn't the lowest, but what is there is beautifully timed. It is among the few metal dome tweeter loudspeakers that do not provoke listening fatigue in any way. The Celestion SL600 produces extremely good imagery with very good depth.

The SL600 adopted the same driver and crossover network but was superior in neutrality, it just sounded wonderful

HiFi journalist Keith Howard had written an article about cabinet construction and suggested that the future was to make a cabinet that was 'vanishingly light and infinitely stiff'. The SL600 was intended to be just that product. 'Journalists were slathering over it,' remembered Form, whose early training as a research chemist had given him a further advantage—enabling him to solve the unwanted 'bell ring' problem in the first-production Aerolam cabinets by replacing the epoxy resin of one joint with a long-life silicone elastomer. 'The SL600 adopted the same driver and crossover network as we had used previously but was superior in neutrality, it just sounded wonderful.'

Acoustic Ribbon Technology
Bank had left the company for two years, returning in 1985 to become research director, and soon after this he set about working on what became known as Acoustic Ribbon Technology (ART). Ribbon loudspeakers were on the more esoteric end of the loudspeaker driver market and were to be seen mainly in higher-end HiFi brands such as the well-regarded Apogee. The Celestion unit had taken almost three years to design, making use of the fledgling Finite Element Analysis process—a technique that had recently been adopted to model loudspeaker behaviour using advanced mathematical methods, which in turn had been facilitated by the increasing availability of desktop computing power.

Ribbon speakers were considered to be exceptionally low in distortion, free of many of the unwanted fizzes, honks and squawks that were apparent from dome tweeters when listened to under laboratory conditions. The resultant ART was a specialised mid-range/treble unit that operated from 900Hz to 20kHz, with a notably high sensitivity of 86dB Sound Pressure Level (SPL). Mounted in a cast aluminium frame, it had an effective radiation area of 10mm x 530mm (the same as a regular 5in speaker).

However, Celestion's ribbon technology took a different approach from others: it was monopole rather than dipole, radiating sound in only one direction, with the sound from the back of the ribbon reflected back off the surfaces in the speaker cabinet, to boost the overall output sensitivity of the driver (low output being one of the weaknesses of ribbon devices in general). It also featured a patent applied for suspension developed by Carl Pinfold, referred to as the RS system, which supported the pure aluminium ribbon using 60 strontium ferrite magnets.

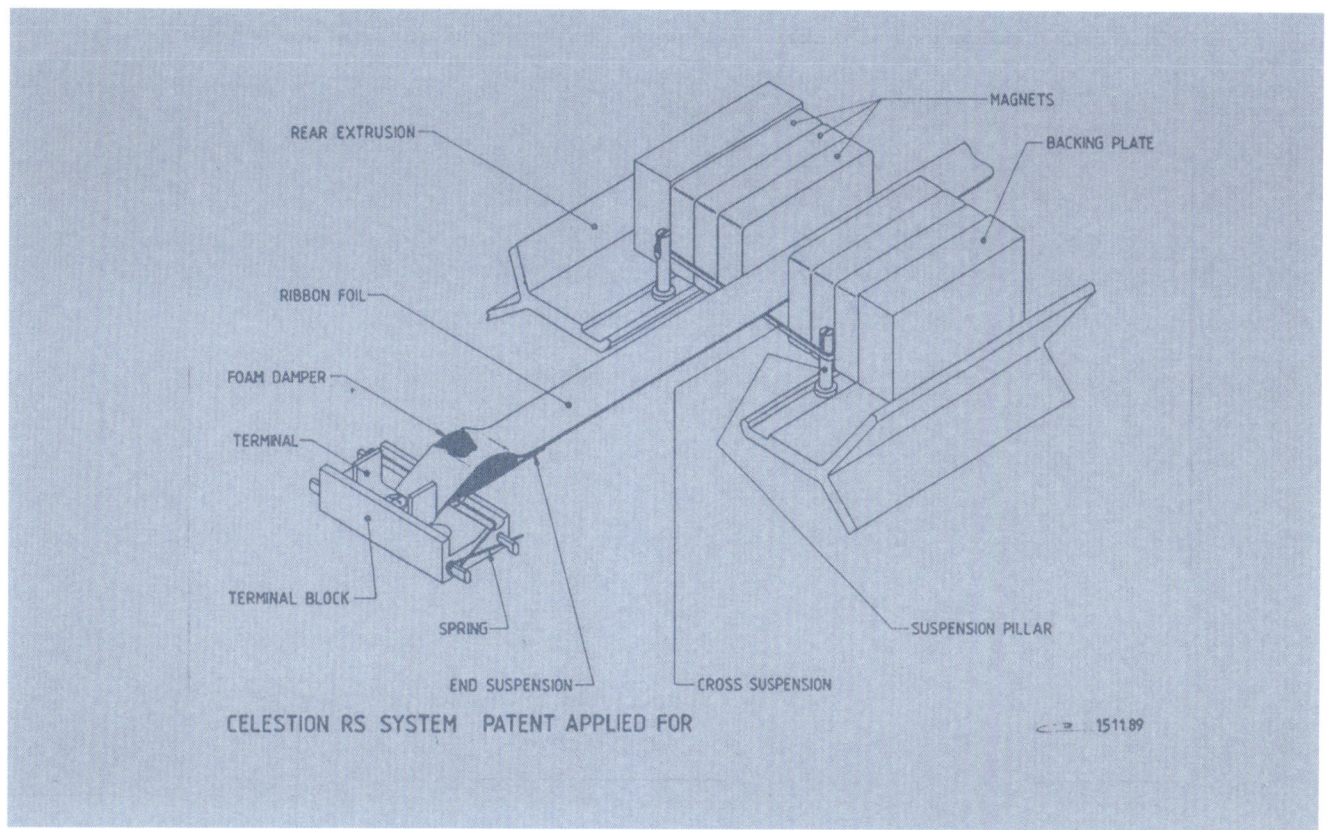

Patent diagram for the the RS System ribbon tweeter suspension.

Sonically and technologically, ribbon transducers were a distance apart from metal dome tweeters, which had become Celestion's heartland, but this represented a big investment for the company as it once again attempted to move further upmarket. ART was initially featured in the Celestion 3000, the first of a small range, which was launched in the autumn of 1989 at the popular Penta HiFi Show in London.

Referred to as ribbon/dynamic hybrid speakers by Celestion, the 3000 and 5000 (identical except for cabinet finish) both featured the ART alongside a traditional cone loudspeaker that handled the mid/bass duties. The more expensive 7000 featured two mid/bass units. As John Atkinson's review of May 1990 concluded,

> Beautifully made and incorporating a well-thought-out ribbon driver, the Celestion 3000 is an impressive design. Its seamless, coloration-free midrange, tight low frequencies, and exceptional transparency are excellent by any standard and are matched by a good sense of dynamics up to SPLs of around 100dB or so. Its unique design will also ensure that more than one listener will be able to hear both a well-balanced sound and a reasonably well-delineated soundstage. With the right kind of music and optimal electronics, a pair of Celestion 3000s will produce a sound that is vividly detailed, uncoloured throughout the midband, and has excellent image depth.

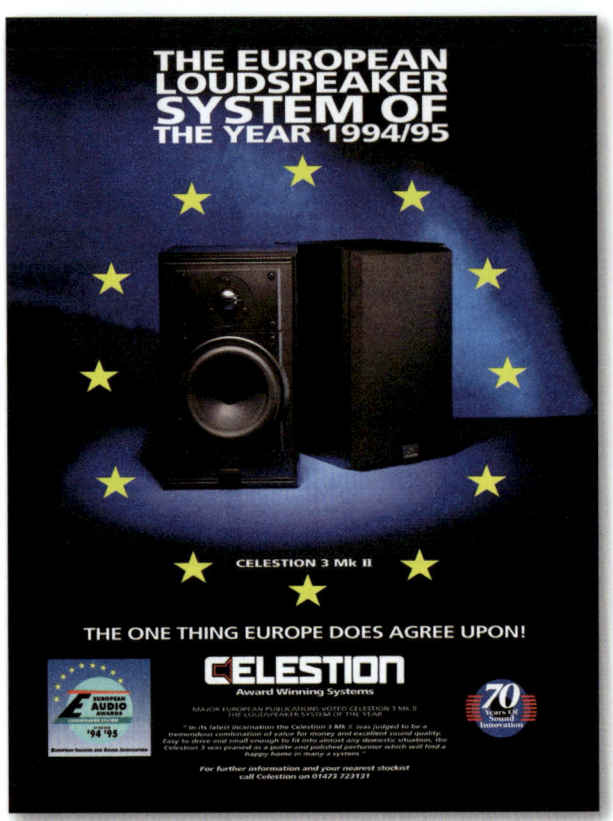

Left: The compact, competitively priced and very popular Celestion 3.

Right: The later Celestion 3 Mk. II version secured 'European Loudspeaker of the Year' for 1994/95.

HiFi for Everyone

In the engine room of Celestion HiFi development, attention had also turned to the soon-to-be award-winning DL Series. Much more in keeping with Celestion's ethos of 'HiFi for everyone', it was a deliberately more cost-effective take on the SL range, with some of the new technologies and ideas ported across. The DL Series was a range of (what were by now considered to be) traditional bookshelf and floor-standing cabinets, originally with a soft plastic dome tweeter, and later benefiting from the design of a new titanium dome tweeter, conceived by Bob Smith, to boost treble output.

With the HiFi programme on something of a roll, what followed was later known as the 'unit' range (which also featured the new titanium dome tweeter); the products were simply designated 1, 3, 5 and 7. Launched at the massive Consumer Electronics Show in 1989, the spearhead was the Celestion 3. Ultra-compact and ultra-competitively priced, at under £100 in the UK and $250 in the United States, the product was designed to a stringent cost target set by Managing Director and newly appointed Chairman Gordon Provan.

The Celestion 3 is one of the most outstanding small speakers I have heard, and an absolute best buy at its price

On his arrival two years previously, Provan had been focused on reorganising the three main international subsidiaries in France, Germany and the United States, in order to reinvigorate export growth. Having completed that mission, he set about repositioning Celestion as an accessible, mass-market HiFi brand for the changing audio landscape of the late 1980s.

The unit range, and the Celestion 3 in particular, was fully intended to take the budget end of the HiFi market by storm and gain both the market share and the additional brand recognition that Provan deemed was absolutely the company's route to greater success. 'Sales of the Celestion 3 are running at twice the rate the company anticipated,' boasted Provan in April 1989, 'and the main objective has been achieved: to improve anyone's budget HiFi system.'

Reviewers were suitably impressed by the overall sound quality and that Celestion were able to incorporate a titanium dome tweeter into such an affordable range. 'The Celestion 3 is one of the most outstanding small speakers I have heard, and an absolute "best buy" at its price,' gushed *HiFi Classic*.

The 90s were looking good for Celestion HiFi.

How the Drinkmaster™ Capsule Became an Aluminium Dome

During his time at Celestion, Ed Form was perpetually beguiled by Bill Strong and his achievements: 'He was a genius and could do things with metal that would amaze you. We eventually came to the point where we needed to make the aluminium domes in production [for the SL products]. We couldn't persuade any of the companies who made them to combine a dome and voice coil former in a one-piece pressing; they all said they didn't think it could be done. I also asked a company that made takeaway trays and they said it should be possible, but they then asked: 'Can the wrinkles be on the dome or in the sidewall?' They just didn't get it.

'I had already asked the production engineering team, who agreed it was possible, because Bill was already making samples for us, but they professed their ignorance of tool design for turning flat aluminium foil into something so 3D and so delicate. Then, when I went back to them with my failure to find a solution, Brian Brame opened his desk drawer and produced several "Drinkmaster" capsules, those flat pressed aluminium pots with an airtight paper disc on top that you buy from a vending machine, peel off the paper seal, pour the granules into your mug and then fill it up from the hot water dispenser.

'Brian pointed out that the pot itself was immaculate, with no trace of wrinkles, but he had also cut a section from one and showed me that the flat rim was actually folded multiple times into a flat sandwich and terminated in an exquisite circular bead about a quarter of a millimetre across, so no one could cut themselves handling it. His opinion was that I should find out who made these little miracles and seek their advice on making our dome. So, I did.

'It turned out that the maker was a division of a giant conglomerate and made all manner of pressed aluminium packaging items. The Drinkmaster pots were produced simultaneously, all day long, on several sets of tooling. The boss of the company was initially reluctant to waste time talking to me, because our limited production requirement was impractical on his high-speed presses, but we arranged a meeting with him and his toolroom manager. I took Bill Strong with me, and although the meeting was excellent, and showed us clearly that these folks could design a tool to do the job, the boss said they could not spare the time.

Form rang Celestion Industries chairman, Daniel Prenn, and asked if he could pull any strings. 'It turned out that the chairman of the conglomerate was a friend of Prenn's—and orders from on high got us a set of tool drawings, plus five days of the toolroom manager's time. We first had to make the tool parts and then the toolroom manager was to spend five days at Ipswich, at our expense, to

Close up of an SL Series aluminium dome.

assemble and commission the tool.'

By late afternoon on a Friday, the tool was just about working. 'The proposed method of making it work was to press a part, see which side it tore, and then bray the opposite side of the ring-die with a lump hammer to distort it very slightly then rinse and repeat. I expressed my thanks to the toolroom manager and off he went home—the plan being that our toolmakers now knew what to do and would finish the job on Monday.'

But Bill Strong was a little distrustful of what had gone on. Form was sitting in his office in the lab long after closing time, when Strong walked in and said, 'Please let me make a new punch and die.' When Form arrived at work the following Monday, Strong had made new parts over the weekend, and was just finishing the re-assembly of the tool. Form recalled: 'A few minutes after I entered the production room where the tool was set up, Bill pressed the button, the tool closed, opened again, and ejected a perfect dome, no hammer required. From that point on, until I left the company, the tool never made a bad pressing.'

12 G12 and the Quest for More Power

As the Swinging Sixties segued into the 1970s, continuing developments in musical instrument amplification and the budding sound reinforcement industry prompted Celestion to establish the Power Range, with the G12 at the forefront, as the go-to 'heavy-duty cone loudspeaker' for high-power applications.

By the end of the 1960s, four flavours of the G12, with four different sizes of ceramic magnet—the G12L (Light: 16oz), G12S (Small: 27oz), G12M (Medium: 35oz) and G12H (Heavy: 50oz)—had been firmly established. Features included a 'Feroba II' magnet system that 'ensured exceptional efficiency and control'. Voice coil leads were 'specially terminated to eliminate the possibility of fracture', with 'all moving parts, housed in a rugged chassis to ensure permanent alignment'.

At this time, they were available with a paper edge—the 'standard' version, which would have been used principally by guitarists—or a cloth edge, allowing for applications that required greater cone excursion, or where the cones simply needed to work harder. Applications would have included keyboard, bass guitar and PA or sound reinforcement systems (such as those produced by WEM).

Also available were versions with metal dust domes and twin cone variants, where a 'whizzer' cone was positioned in the middle of the main cone for the purpose of extending the output frequency response to 10kHz and beyond: in some applications, this would have negated the need for a tweeter altogether.

As the 1970s progressed, it was the G12M and G12H models that remained at the forefront of guitar speaker tone. Thanks to their adoption by the majority of British amplifier firms, these speakers were cemented firmly in place as the voice of rock 'n' roll for all those seeking the 'British tone', helping Celestion to remain the foremost name in guitar speakers. As a result, dozens more soon-to-be-legendary players joined the pantheon of Celestion players during the 70s, a small sample of whom included Brian May of Queen, Joe Perry and Brad Whitford of Aerosmith, Angus and Malcolm Young of AC/DC and Edward Van Halen.

Opposite: AC/DC were one of the many Celestion-loaded bands to emerge in the 1970s.

Amps became more powerful and overdrive more savage, and there was an increasing need for speakers of greater power capacity

More than anything else, however, it was the successful and enduring relationship with Marshall that served as the engine for new guitar speaker innovation. As the decade wore on, amps became more powerful and overdrive more savage, and there was an increasing need for speakers of greater power capacity. A speaker's power handling is determined to a large extent by the heat-resistant properties of its component parts—in particular, the former around which the voice coil is wound. All the electrical input energy passes through the coil and a great amount of heat is generated (the loudspeaker is not very efficient at converting electrical energy into sound, but, as it needs to go somewhere, the energy is instead dissipated as heat).

The earliest ceramic magnet speakers used kraft paper as the main body of the coil former, but its ability to cope with heat was negligible. Fairly early into the 1970s, this led to the adoption of alternative materials, notably fibreglass, to increase the heat-defying properties of the voice coil, imbuing the speaker with additional power-handling capability so it could be used with amplifiers of greater output (as well as altering the tonality). Thus, the race for more power reached another stage.

The G12/50, first supplied in 1974 with its distinctive yellow chassis and black rear can, was probably the first guitar speaker to feature a fibreglass voice coil former, determining its 50W power rating. It was also the first G12 speaker documented as featuring a black (rather than a green) can. This was followed by the G12/75, which later led the way to a full range of new-style G Series speakers that superseded the former G12M, G12H, et al.

Conceived in 1978 and incorporating 8in, 10in, 12in and 15in models, this range catered for a wide range of applications with a huge variety of construction types, including paper and cloth surrounds, twin cones and aluminium dust domes. Of these, it was the G12-65 that became the standout guitar speaker model and the new go-to speaker for the Marshall 4x12 cabinet.

This race for power wasn't just about guitar speakers. At the beginning of the 1970s, Marshall was keen to move into the rapidly expanding sound system market, which at the time was still dominated, in Europe, by WEM. To do this, Marshall financed a system for the Weeley Festival (Clacton, Essex) in 1971. The system was a joint venture between Nigel Olliff, later of Martin

A stack of G12 cones prepared for assembly.

Audio and BSS, Nick Gilbey, Jo Browne, who set up Marshall Equipment Hire (which eventually became TASCO) and Ken Flegg, who had been involved in the design of Marshall's first 100W guitar amplifier. All bar one of the festival acts were mixed through the Marshall branded cabinets all of which used Celestion cone speakers and horn drivers.

As early as 1972, Celestion had conceived the Powercel range, with 12in and 15in cast aluminium frame bass speakers (coloured yellow and fitted with black rear cans.) Intended to be more 'heavy duty' than cloth-edged versions of the G12, these too were initially built in partnership with Marshall and were used with the cabinet that accompanied the Marshall 'Supa Bass' PA cabinet. A 15in Powercel was also featured in Marshall's 2052 lead-organ cabinet, to produce the extended low frequencies that would have been required. Perhaps because of their prior association with Marshall, London-based TASCO remained major users of the Powercel 15 in their 4560 bass bins and wedge monitors, which were used by The Who on multiple tours.

At Foxhall Road, G12s, and the smaller units designated G8 and G10, were assembled on the 'Forth' production line (named after the Scottish river), which was also colloquially known as the 'Power Line'. Larger units such as the G15 and G18, and the Powercel speakers, which by nature and application were larger and a lot heftier, were built on a separate line referred to simply as 'Heavy Power'.

Power speaker manufacturing at Foxhall Road.

Celestion continued to supply Marshall with speakers for its PA cabinet designs throughout the 1970s. The 50W system included a pair of 2047 columns, which were front-loaded horn units, each containing one 10in and one 12in speaker. The 100W system contained a pair of 2043 columns, also with front-loaded horn units, each containing two 10in and two 12in speakers. The compact 2097 housed eight 8in speakers.

By the early 1980s, power handling had become somewhat of an arms race, and the introduction of yet another revised G Series opened up even greater power-handling capability, thanks to the incorporation of modern, lightweight polymers into the voice coils, like polyimide (trademarked as Kapton® by DuPont). With drivers from 5in all the way up to 18in, the early 80s G Series was the go-to product for all musical applications—truly the general-purpose speaker that the G had always stood for.

The G12M-70 (aka the Modern Lead) became the Marshall 4x12 speaker of choice for a brief period until the custom-designed G12T-75, with its larger 1¾in voice coil, tighter bass and beefier sound, appeared in late 1984 and swept away everything that came before it, as the undisputed Heavy Metal speaker.

1986's B18-1000 was reportedly the first-ever 1000W bass driver

It wasn't just G Series speakers that were gaining in power-handling capability. The Heavy Power line continued to produce PA speakers whose capacity increased in lockstep with the needs of ever-expanding sound systems. 1986's B18-1000 was reportedly the first-ever 1000W bass driver and specified for use (albeit in customised form) with Martin Audio's BSX, RS800, VRS800 and later VRS1000 cabinets.

Around the same time the T75 was beginning to dominate, a new generation of speaker was introduced with the intention of offering a distinct alternative to the G12: the Sidewinder. With a model suitable for all amplified musical instruments (lead guitar, bass and keyboards, at any rate), it was, according to the brochure, 'developed to bring modern music and acoustic science firmly together'.

Its development relied on the use of the increasingly ubiquitous laser interferometry to analyse speaker behaviour, creating a new style of speaker that brought together the 'best of British and American speaker design'. It employed an edge-wound (or side-wound, hence the name Sidewinder) voice coil, where the voice coil's wire has a rectangular cross section instead of the more conventional round cross section used by the G12 family, for closer winding that increased output sensitivity.

According to the brochure for the Sidewinder,

> Until now, guitarists have had a choice. Either Celestion's classic G Series with their rich, gutsy rock 'n roll performance, or American made units with edge-wound voice coils and a cleaner, country feel to their sound. Now there is no choice: the new S12-150 does it all! From open bluesy riffs and country music to the deep dark snarl of Heavy Metal.

The purpose of the Sidewinder was unmistakable and the message was clear: here was a Celestion guitar speaker that, at a time when the company was hungry for export success in the United States, looked across the Atlantic and to the US speakers, whose sound the company had so deliberately avoided up until this point.

Such a clear divergence from the norm was always going to divide the crowd, which is what the Sidewinder did, and it had as many detractors as it did fans. However, as the 1980s progressed, another change was in the wind. The market for guitar speakers began to take a turn for the retro after more than a decade of modernisation, and the sound of the 60s was back in vogue.

The Vintage 30

Creating the signature sound of 60s guitar was one thing—emulating it nearly a quarter of a century later, when times (materials, suppliers and manufacturing processes) had moved on, was quite another. But when Marshall came calling once more, Celestion was required to respond with a new speaker that had a sound that was comfortably close to the old sound.

Shortly before Form left the company in 1986, he and Development Engineer Ian White were presented with a pre-1971 G12M, which they cut open, Form remembers, 'and discovered a rolled-in compliance-groove in the coil former, something I had no idea ever existed'. White, who at the time had just arrived at the company, remembered the mandate. 'However, it wasn't a pre-1971 G12M it was an Alnico Blue, Marshall was putting pressure on us to get the vintage tone. So, Ed gave me this as a project.' The result was to be a custom-labelled speaker simply called 'the Vintage'.

At this time, Celestion were liaising with Marshall's amp specialist and chief engineer, Steve Grindrod, who believed that 'ceramic speakers did not sound as good as alnico speakers'.

The Vintage was originally intended purely for the Studio 15 combo (Marshall model 4001), and, in the Marshall man's view, '[It] was not intended to be a G12H-30 type but rather a modern-day alnico. The price of cobalt was outrageous due to the cold war, hence the use of the ceramic magnet. We found that the 'H' magnet came closer to the alnico in terms of measurable performance, and a new coil assembly was designed using modern materials having similar mass and physical properties to the paper coil former of the alnico 15W, but with a much higher flammability rating; hence the Vintage's 70W power handling.'

'Unfortunately, the original cone was not available, so we used the next nearest and added some quite severe doping. The result was a very fine loudspeaker, not quite but very close to the sound of an alnico with a fatter midrange. Since its release date coincided with increased demand for classic loudspeakers such as the G12H-30, it was marketed by Celestion as the "Vintage 30".'

After a grace period, where Marshall was granted exclusivity, the speaker was indeed restyled as a standard product (although with a more conservative power rating of 60W for extra safety), though not everyone at Celestion would agree with Grindrod over the origin of the name. Some suggest that the '30' in Vintage 30 refers to 30cm (the metric equivalent of 12 inches, the diameter of the speaker). And, with the development of the 10in-diameter Vintage 10 and the 8in-diameter Vintage 8 following just a few years later, it's a very plausible explanation. The definitive answer, alas, is lost in the mists of time.

The development of the speaker was certainly pioneering. 'The G12 Alnico Blue was interesting—it was quite different from a lot of the drivers at that time,' continued White. 'At the time [the late 70s/early 80s], most guitar speakers had Kapton or fibreglass formers, but this had

Left: The Vintage 30 guitar speaker.

Right: Vintage 30 advertising.

a cartridge paper former, which was flimsy. It was important because it allowed the cone to "waggle" [producing lots of desirable harmonics]—so it was quite significant. I didn't want to use cartridge paper because of the risk of fire. There was an alternative material we hadn't tried but had been used previously in HiFi speakers. So, I started making prototypes with this instead.' This was a 'meta-aramid' material not unlike Kevlar® but which had stiffness properties more similar to paper, although with a much higher heat dissipation characteristic. It ended up enabling the speaker's significantly higher power rating.

The laser system was used to analyse the behaviour of the original Blue cone and, more than anything else, this was the key resource that led to the development of the Vintage 30. 'It was especially good at looking at cones,' remembered White. 'We used that data to form a precise model of the vintage speaker's characteristics.' A crucial step in re-formulating the cone.

'The speaker really took off and I felt very proud.'

Mysteries of the G12 Revealed

When Is a Greenback not a Greenback?

Very soon after their inception, the original ceramic magnet G12s were supplied with a green cover to conceal the magnet. This was presumably for aesthetic and practical purposes, as the iron-based ceramic that was used to make the magnets was brittle and easily chipped. At the time, the company was producing a variety of public address speakers made in 'industrial' colour schemes, such as the appealingly named 'drab olive', so it was only natural that they would default to something like this when producing new hardware parts, and thus the 'Greenback' was born. The can itself was produced in various sizes, dependent on the speaker's magnet size (S = Small, L = Light, M = Medium and H = Heavy), although it was the G12M and G12H that proved the most enduring.

The Greenback had originated in the mid-1960s, but by the mid-70s Celestion was experimenting with alternative can colours, with black, cream and grey all being used at one time or another. Of all of these variations, the black can has the greatest provenance, having initially been used on the G12/50 speaker made for Marshall in 1974.

The standard Celestion explanation of the variation in can colour has long been that there were periods of time when the company simply ran out of green cans and couldn't get any more quickly enough. Although plausible, as we know that black cans were in use elsewhere, it's not particularly satisfying.

Another suggestion was that the different coloured can denoted when a different cone supplier had been used. For example, long time curator of vintage Celestion guitar speakers Brian Harding suggests that the black can corresponded with the adoption of Kurt Mueller branded cones. He also speculates that grey cans were adopted from mid-1973 to early 1974, while cream cans were deployed from early 1974 to early 1975, potentially at the same time as the seldom-seen 98700 cone, about which very little is known, was in use. Black cans were used from 1975 to 1980, until around the time the higher-power range of G Series speakers took over and the use of rear cans was abandoned altogether.

The Legend of Pulsonic Cones

Perhaps because of the longevity of the speakers themselves, along with the affection and enthusiasm that music fans have for the players that played through the G12s and the raw sounds the speakers made, every detail of the production of these special speakers has since come under the microscope.

As the saying goes, 'the tone is in the cone', so inevitably the cones have undergone more

The first G12H came with no can. G12s were later supplied with green cans as standard and for a while grey, cream and black cans were used.

scrutiny than anything else by players and collectors alike. When talking Celestion cones, the magic word is 'Pulsonic'. It seems that the Pulsonic cone is the holy grail of speaker tone: offering a sweet, fluid sound that can at once propel you into the lofty presence of Hendrix, Clapton, Beck and Page, and all the others in that pantheon of guitar gods assembled in the late 1960s, whose sound is for evermore to be an inspiration to those who wield a guitar.

But who or what actually IS Pulsonic? It's a complicated question.

Brian Harding is confident he has traced the earliest usage of Pulsonic cones by Celestion to 1962, by careful examination of the codes that are stamped on the outer side of the membrane. Pulsonic, according to Harding's research, was the cone of choice for G12 manufacture up until 1975, although legend has it that Pulsonic's premises was destroyed by fire in 1973, the implication being that the secret to the company's otherworldly-sounding products also went up in flames. Afterwards, there was a gradual adoption of cones made by a company named Kurt Mueller, which, although of superb quality, could never quite match up to the Pulsonic tonal standard.

Before 1962 and for a period of a couple of years during the transition between Pulsonic and Mueller, G12 cones stamped with the letters 'RIC' were reportedly used. It is thought that these cones were made in-house at Celestion.

On deeper investigation, 'Pulsonic Ultrasonics' was registered as a company in

Celestion itself had plant for manufacturing cones in the Thames Ditton factory

1970, although the same company started life as C.T. Chapman (Reproducers) Ltd in 1950, founded by engineer Cecil Chapman. Renamed Chapman Ultrasonics in 1960, the company was acquired by Derritron, a manufacturer of 'electrodynamic products' (transducers by another name), and subsequently renamed 'Derritron Ultrasonics' in 1963.

The change to Pulsonic came a few years later, and what is also clear from the published company documentation is that Pulsonic did indeed appear to undergo some seismic transformation around 1973 which ultimately necessitated a change of premises. Exactly what happened is now long forgotten, but according to archive information the company appeared to be fully back in business soon afterwards and by 1978 was trading as Kurt Mueller (UK) Ltd. What's also known is that Dr Kurt Mueller himself had been involved in the Pulsonic company from its establishment (in 1970), perhaps as a way of expanding to create a new UK factory for his German company.

Whether cones that were made prior to 1970 were known as Pulsonic or something different is unclear, but it seems very likely to have been the same company that made these cones throughout the majority of the 1960s, whatever name they happened to be trading under at the time. It's a testament to both the enduring nature of the cone design and the specialist skill of the cone manufacturer that, to this day (at time of writing), Kurt Mueller (UK) Ltd is still supplying cones for modern-day guitar speakers, including the G12Ms and G12Hs.

Celestion itself had plant for manufacturing cones in the Thames Ditton factory, and it would have been natural that those cones made for the earliest alnico G12 guitar speakers would have been 'in-house'. One might speculate that the demand for G12s grew to such a point that it became necessary to outsource production in order to keep up with demand. Likewise, if the production capability of Pulsonic had fallen back between 1973 and 1977/1978, it makes perfect sense that Celestion-manufactured cones would have been used to make up any shortfall.

In-house cone manufacture would almost certainly have come to an end by the close of 1975, once the company had finally exited the Thames Ditton factory completely, since Celestion, according to engineer Bob Smith, had already decided against relocating the cone press to the Foxhall Road factory.

As for the meaning of the code 'RIC' that was stamped on the in-house G12 cone? That, along with the other cone codes, including JIL, ROB and NIB, all remain mysteries that continue to elude detection.

Left: Left: A late 1960s 'Pre-Rola', Celestion Ltd-labelled G12H.

Right: An early 1970s Rola Celestion-labelled G12H.

The Pre-Rola Question

The earliest ceramic G12s, from their origin in 1965, were all given Celestion-branded labels. This remained the case until April 1971, from when the earliest example of a Rola Celestion labelled G12 has been found. This is a transition that makes perfect sense. It happened soon after the formation of Celestion Industries (and the subsequent move to Ipswich), but still left long enough for the manufacturing in the new plant to become fully established. It also illustrates that the Rola Celestion brand had come back into favour once more, which is also evidenced by much of the associated media and company documentation of the time, likely as a way of differentiating the audio division of the newly minted Celestion Industries plc from the textiles division.

The new Rola Celestion label contained the updated Celestion logo, which can be seen in subtly different guises throughout the 1970s, adapted for the different speaker types. It also contains the Foxhall Road, Ipswich, location (supplanting Thames Ditton, Surrey) at the bottom of the label. The more eagle-eyed may also notice that the label swaps c/s (cycles per second) for the more scientifically standardised, but prosaic, Hz (Hertz).

As we now know, British Rola acquired Celestion in 1946 and, despite bankruptcy coming soon after, the name Rola Celestion endured as a trademark, for the most part due to the patronage of its chairman, Daniel Prenn. As such, Pre- and Post-Rola isn't much of a historical signifier when it comes to guitar speaker products. The label changes happened gradually, however, and examples of 'Pre-Rola' speakers have been found dated as late as 1975.

13 Sound Reinforcement Systems

It is arguable whether Celestion would ever have fully entered the world of complete sound reinforcement systems in 1987 had a gentleman by the name of Roger Williams, who ran local contract vehicle hire business Wilhire, not come knocking at Celestion's door five years earlier.

The proposal was to design and build a series of heavy bespoke cabinets, for large-scale sound reinforcement, the first time the company would have attempted a system of this size. The idea itself was a risky undertaking for Celestion, and to do it at the behest of a man with no previous history in the entertainment business who had the notion of opening something called a 'roller disco'? The whole endeavour seemed more than a little bit crazy.

The disco scene in its many forms was just about at its apex in 1982, dovetailing with a new era of sound reproduction, with powerful 'public address' speakers for sound reinforcement becoming more available, affordable and scalable. So it was that roller discos, in which servers and clientele alike would dance, generally on quad skates, became a successful and popular, if short-lived, trend within the disco genre.

It was in this environment that Roger Williams conceived 'Rollerbury' in Bury St Edmunds, Suffolk—what was to become the largest roller disco in the UK, just 30 miles from Celestion HQ in Ipswich. And he wanted Celestion to produce a flown bespoke sound system for the venue.

The approach had been radical to say the least. Celestion was able to make use of raw frame loudspeakers that were already part of the 'Power' range, but so many drivers were loaded into their cabinet designs that the boxes finished up at a third of a ton each. Paul Fidlin was Celestion's main engineer on the project and produced the crossover networks, while highly respected cabinet builder Acoustic Sound Systems in Southend built the enclosures. Assembly took place at Foxhall Road.

Opposite: A cutaway view of the SR1 with its many features.

The Rollerbury roller disco, Bury St. Edmunds.

Inset: Advertising for the Rollerbury system.

With weight-bearing in mind, Celestion hired a local expert rigger to install these beasts, complete with the mandatory safety chains. This solution for Rollerbury was a total one-off—with 6in x 15in bass drivers, 3in x 12in mid-range, plus Celestion compression drivers with a radial horn and an array of the popular HF50 and RTT50 horn tweeters in each of the four clusters.

This coincided with Celestion announcing its new wide dispersion, short-throw AL7 and AL12 slant-plate lenses, designed to control the HF dispersion of those HF50 and RTT50s. Acting as angle-dependent delay lines, they gave a broad, flat-fronted coverage pattern, ideal for short-throw use. Aimed at the new breed of DIY enthusiasts, both were simple to fit and of very robust, all-metal construction.

Once complete, this custom installation system was featured in a full-page ad appearing in the nightclub trade journal *Disco International* under the heading POWER TO THE PEOPLE, dedicated entirely to the muscularity of the sound system.

Celestion decided this 'made to measure' sound should become available to all DIY enthusiasts via a new *Cabinet Handbook*. This contained all the information necessary to build a custom

Left: A specification sheet for the P1 'professional loudspeaker system'.

Right: Systems advertising in *Disco International* magazine.

sound system perfectly suited to requirements, enabling audio fans and hobbyists to deliver on stage the sound professional musicians had come to expect from Celestion. Clearly this had been a lightbulb moment at Foxhall Road.

The company had already flirted with the idea of sound reinforcement speakers, having offered several column speaker models during the 1960s, and by 1980 it had marketed the 500W P1 system as a compact and roadworthy touring system. However, it was the successful, large-scale system built for Rollerbury that had proven to be the real catalyst, and by the mid-80s the concept of Celestion Systems had begun to be taken more seriously.

The resultant design was for a cabinet that contained 2in x 8in drivers, which incorporated a decoupled coil and rigid hemispheric concentric centre dome for extended HF response. The cabinet would come complete with a controller/limiter that dynamically adjusted the low-frequency response to optimise bass levels and protect the drive units from both excessive LF input and thermal overload. This was the SR1, the first product in the long-running and revolutionary SR Series.

SR1s provided sound reinforcement for the Snape Maltings theatre, Suffolk.

Ian White, who by this time was head of engineering, later reflected on the SR Series with enormous satisfaction, having been acutely conscious of the huge development costs and processes necessary to deliver the glass-fibre-reinforced polypropylene enclosure: 'It was quite a risky thing for us to do; the injection moulding tools for the SR1 cabinet, let alone the SR3 and SR Compact that came later, were horrendously expensive at £50,000.'

'The speaker was in two halves—both injection-moulded, with the carcass at the back. We had to develop a way of fixing the two together other than with conventional glue, because some of the applications would see a lot of rough and tumble. I think we were led in the direction of a welding technique used to bond lorry or car bumpers together.'

Thankfully the process was successful, the whole thing fused together well and the cabinets remained rock solid: 'We were good at producing hard domes and using laser interferometry we could make stiff domes so we married up the large dome to do the HF in the centre of an 8in cone for bass and mid—which worked extremely well.'

With the two 8in concentric dome radiators, they introduced a specially designed cone surround for wide dispersion and reduced distortion which was later named Flexirol™. It gave a very good response, being a very neat way of controlling excursion at high levels. Paul Cork, who would later take over from White as head of engineering, explained the rationale: 'If you look at those woofers you can see a series of 'dimples' on the inner roll of the surround. As the cone moves forward or back the cloth stretches and without the dimples the cloth comes to a hard stop (which causes excessive distortion). But with Flexirol the surround gathers the cloth from the dimples which allows a slower braking of the cone, and hence much less distortion.'

The SR1 was designed to be mounted straight onto a tripod with the large SR2 1x18in 1000W sub-bass at the bottom. Celestion was soon offering superior bass bins in the shape of the SR4, SR6 and SR8—the first two released in 1990 and the SR8 in 1991 and based on another new Celestion technology: 'Paraflow'. The 2in x 10in bass columns came in two versions, for the SR4 and SR6 respectively, while a 2in x 15in was fitted into the larger SR8.

Paraflow was essentially a novel band-pass method of cabinet loading. According to Ian White, 'with the 2 x 10s, for example, each 10in was in a reflex enclosure, with ports arranged to be concentric. What it actually provided was a natural band-pass. It was extremely efficient over 50–200Hz: the response came in at 50–60Hz and went flat to 200Hz and then died, which was ideal for a bass bin and that was all done through this parallel reflex porting.'

The same was true of the 15in drivers loaded into the SR8, where one of the drivers would be electronically reversed. The flow of sound from each individual chamber was arranged to exit in phase and produce a natural band-pass response and SPL of 126dB. The SR8, in particular, had been a real triumph, believed White.

The system's accompanying SRC1 'smart' controller, meanwhile, managed to lift the falling high-frequency response of the dome and provide some boost at the top and bottom end, with a novel protection device on the bottom end. Celestion had been among the first ever to use an EQ processor as an integral part of a system, which would cut the bass and extreme LF depending on level. The sensors that went back to the cabinet monitored the excursion and any thermal problems, as it was essential the voice coil was kept cool.

The SR Series entered the marketplace at the beginning of 1987, making its global debut to widespread acclaim at Europe's predominant spring trade fair for musical instruments, the Musikmesse in Frankfurt. Subsequently, it sold well, not only to bands (many experiencing a proper PA system for the first time), but also to the burgeoning conference centre and retail markets.

In fact, Robbie Gladwell, an English rock and blues guitar player—who at the time was a Gibson guitar demonstrator—was the man responsible for forming a house band to demonstrate the system at key expos throughout Europe, and who recalled the origins of the project.

Having first crossed paths with Celestion in 1987 at the Paris Music Trade Fair, he remembered: 'I was doing a guitar demo for Gibson and I noticed a couple of black-suited guys and one of them started talking to me about a new system they were working on called the SR Series. As I was a

An SR installation at Ronnie Scott's jazz club in London.

musician with a scientific background, I was asked whether I would assess the design features and do a listening test to see if they were fit for purpose.'

Introduced to Ian White, and asked for his thoughts on mobile PA systems, he realised that his design template more or less coincided with that of the manufacturer: 'My view was that it had to be a portable, lightweight system, which is exactly what they were working on.'

By the 1988 Frankfurt Musikmesse, a new demo band was up and running, the almost-legendary Spank The Badger—whose name was based on a *News of the World* headline that caught Gladwell's eye as he passed a newspaper hoarding at London's Liverpool Street station. Aside from the guitarist wielding his Gibson M-III, the band featured his friend Paul Airey (who would later join the Celestion sales team) on keyboard, Tony Muschamp on bass and Francis Seriau (later a renowned music educator) on drums.

Reflecting on the idea of live bands showcasing product at trade fairs, Gladwell states, 'We were ahead of the game in that respect. And it wasn't only about Celestion, I was bringing Gibson along. We won the best band at Frankfurt three years running—always playing through a pair of SR1s.'

The reception people gave us with that system was fantastic

Gladwell clearly evangelised about the SR1: 'I loved the sound—it was one of the first speakers used with stacked systems, before the line array became popular. We would use four SRs along with the big SR2 bass bin. The only downfall for me was that with a separate amplifier and controller for live band music it presented an awkward lift for pub gigs. In fact, for 14 years I was probably one of the only people going out with a great-sounding Celestion PA in a duo situation. But it was well worth it because the reception people gave us with that system was fantastic. For the power it could deliver from that size it was remarkable.'

By the time the band arrived at the 1992 PLASA Light & Sound Show at London's Earls Court— for what was possibly their last performance—keyboardist Rick Wakeman had rocked up and sat in with Airey and Gladwell. Now developing a real cutting edge, the company was launching a simultaneous two-pronged attack on both live sound and the burgeoning nightclub industry.

Rick Wakeman well remembered the Celestion set-up: 'I had been searching for years for the ideal monitoring system for my keyboards, and Celestion heard about this and contacted me. I saw the system at the Frankfurt Music Fair, which I went to every year and where Celestion were a major player. I tried it out with my keyboards and loved the sound ... and they looked great as well for the stage. And that was the start of a long and enjoyable association.'

On a more whimsical note, he also remembers putting a band together for a human circus called Cirque Surreal, for which he supplied all the PA: 'I bought the system myself it was all Celestion speakers and monitors and a Crest mixing desk. It was very reliable, considerably more than the circus, which collapsed before I got paid!' Wakeman continues to carry 'a ton' of Celestion speakers which he has to this day.

Other early users of the SR Series had spread from the influential Ronelles nightclub in Cambridge, England, with its cutting-edge tech at the time, to the China Club in Miami and various locations in between. Pop rockers T'Pau, reggae stalwarts Aswad, comedian Ben Elton and the trad jazz Tommy Dorsey Band were counted among high-profile live users. The SR Series even found itself aboard Cunard's prestigious ocean liner, Queen Elizabeth 2.

Continuing the royal theme, the SR was also used in the household of the much-loved Queen Elizabeth the Queen Mother. The occasion was the 50th anniversary of the George Cross Award in 1990, which took place at her residence, Clarence House. According to Technical Support Manager Peter Ambrose, Celestion supplied the PA system.

> By combining its HiFi capabilities with advanced PA driver know-how, Celestion could bring a credible and effective product range to the sound reinforcement market

The initial inquiry had come through Paul Airey, who before joining Celestion had been running the Piano Centre in Southend, Essex, and had been contacted to supply the keyboard. For the sound system he turned to Ambrose: 'The piano was delivered and we turned up. There was no real security that we could see save for a spaniel, which turned out to be the bomb squad sniffer dog.'

'We looked at the wiring and the first socket I plugged into sparks came out—they were still using the old-fashioned round pin plug system. I located a second socket but this meant running the cable on the floor in the function room directly in front of the doorway. Paul and I stood either side of the door and we had to tell everyone to "mind their step". This was because most of them were in their 80s, and as soon as they looked down, they tended to trip!'

The event's entertainment was to be supplied by old-time singer and wartime 'Forces' Sweetheart' Vera Lynn, with the piano provided for her accompanist. Ambrose reminisced: 'I remember when she sang 'Roll Out The Barrel', Paul and I had to catch a few of the guests by the arm as they stumbled around the room. The 90-year-old Queen Mother raised her arms to dance and purposely elbowed the two people either side of her—one of whom was a bishop—showing her insistence that everyone had to join in, as she herself did. Once the singing had finished the Queen Mother got up and left. We were told she had gone to watch the horse racing and then have a snooze!'

The SR Series remained an important part of Celestion Systems through the following decade, with a MkII version launching in 1994. Most importantly, it proved the concept that, by combining its HiFi capabilities with advanced PA driver know-how, Celestion could bring a credible and effective product range to the sound reinforcement market. With that realisation, R&D efforts intensified into the 1990s and several more successful system ranges followed.

However, SR had not always proved popular during its development phase, as Paul Cork later reflected: 'Tests would go on for weeks after development: we used the theme tune from *Shaft*. It was not the original [Isaac Hayes] version, but a disco version that was constantly looping in the power test room. As the technician, I had to check it out and look at the results.'

One can only imagine his slow descent into madness.

Opposite: The SR Series even found itself aboard Cunard's prestigious ocean liner, Queen Elizabeth 2.

14 Celestion International

Since the very beginnings of Celestion, international expansion had been part of the business plan. Very early in the company's history there had been a fully fledged (although short-lived) subsidiary in France (Constable-Célestion), and there are records indicating the existence of a Manhattan showroom around the same time; so straight away export was part of Celestion's DNA, as the company shipped product all around the pre-war British Empire. Financial reports following the coming together of Rola and Celestion talked enthusiastically about dominance of the overseas market, and later, under the umbrella of Truvox, came overseas distribution in the Middle East.

The 1960s had been a decade of prosperity for Celestion—success had continued to come through the sales of components: speakers for myriad industrial and public address applications, and of course the fledgling G12 range that created a new market for guitar speakers, catapulting the brand firmly into the mindset of music makers on both sides of the Atlantic. However, it was the rapid rise of Celestion HiFi that, in the decade following the introduction of the first Ditton model, had 're-orientated the company', as managing director at the time John Church had written in 1974: 'Rola Celestion had become a consumer product company, almost before the company itself recognised it.'

Celestion was beginning to see itself more as a builder of products for sale to home consumers than as a component manufacturer, which it had solidly become over the preceding 25 years. Church promised 'professional marketing and promotion for an international market' and a 'worldwide service and distribution objective'. Once again, the mission was clear: worldwide expansion.

By 1975, Canada-based distributor Rocelco was celebrating its tenth year of association with the company. Having successfully established itself with the Ditton and newly launched UL HiFi ranges in Canada during that time, the company offered the opportunity to distribute throughout the United States 'in what [was] potentially an extremely good market for Celestion products', according to the announcement.

Opposite: Drummer Bill Bruford adopted the SR System, helping to boost sales internationally.

This year also saw the consolidation of a relationship with Narikawa Limited, which was to become Celestion's agent in Japan, an attractive market that was enthusiastic about the 'Celestion sound'. Moving further into the Pacific region, M&G Hoskins was appointed exclusive distributor for Australia and New Zealand by May 1976—again with HiFi products at the forefront.

1977 saw the official announcement of the creation of three Celestion subsidiaries. The French operation, Celestion SARL, had been created the previous year after a takeover of the formerly Paris-based agent Universal Electronics, headed by Gerald Kaas and Blanche Baum, who stayed on to steer the new company. Over in West Germany, Celestion Industries GmbH had also been set up in 1976, headed by Klaus-Dieter Wolf. It was based just east of Saarbrucken, with a warehouse which supplied a large network of regional dealers. The German subsidiary also contained its own servicing facilities; in testimony to that fact, in 1978 Celestion Germany had been voted number one for service by an influential HiFi publication.

However, the biggest prize would be the United States—a move which came into play in February 1977. Sensing the already strong sales impetus from expanding into the United States, Chairman Daniel Prenn sent out the following memo on 20 January 1977: 'In order to develop the successful penetration of the US market by our Canadian distributor, Rocelco Inc of Montreal, we are pleased to announce the formation and opening of our wholly owned USA subsidiary Celestion Industries Inc., with offices and warehouse in Holliston, Massachusetts.'

Whether because they hadn't been successful enough, or perhaps because they were too successful, this seemed somewhat of an about-turn in their relationship with Rocelco. To its credit, however, the Canadian company continued its long-standing and successful marketing, sales and service of Celestion loudspeakers in Canada.

Holliston would soon loom large in the fortunes of Celestion, becoming the company's first *bona fide* US base. Operations commenced on 15 February under a newly appointed executive president: Fellow of the Audio Engineering Society, one time broadcaster and audio industry stalwart John J. Bubbers.

Also joining the board of directors of Celestion, Inc., was John Church, thus making way for Colin Aldridge to take up the managing director's chair back in the UK, the first of a raft of managerial changes that saw Rola Celestion gear up for the continuing push into HiFi and the international consumer marketplace. By 1979 the company's increasingly international profile led to the adoption of the name 'Celestion International' as the identity for the parent company and all major overseas subsidiaries.

By June 1982 William 'Bill' McGrane was listed in *Billboard* magazine as Celestion's North American Sales Manager. Despite the big push for HiFi, McGrane had identified a market demand

Left: John Church.
Centre: John J. Bubbers.
Right: Gordon Provan.

in the United States for the Celestion guitar speaker sound not just from amplifier builders but from the guitar players themselves. He believed it could be satisfied by making Celestion guitar speakers available as a retail product; his role in establishing and creating a demand for the Celestion sound with music stores at that time has since been described as pivotal.

According to Ed Form, 'It was Bill who first got us into selling guitar loudspeakers into the retail trade, places like the Guitar Center in New York. We sold them as boxed retail packs for people who wanted to build their own guitar cabinets and slowly began to create the legend. We shipped them by the ton in 20ft and 40ft containers as guitarists sought to recapture the old sound [of the 60s and early 70s] … everything they read about in the adverts. We were selling a dream, aiming to get back to the classic British rock sound—and it was McGrane who showed us we could sell loads.'

Bill McGrane's strategy had been, by some measure, a success, as evidenced by the fact that by 1986 he had risen to the position of CEO of the entire US operation. However, this had not come without a cost. Following McGrane's departure two years later, Peter Wellikoff took over the mantle, and his immediate priority was to manage cost control after a period of what had been described as cavalier spending. 'I inherited a lot of decisions which weren't fiscally the most responsible,' he believed. 'Bill was spending a lot of money to achieve his aims. His emphasis was more on the distribution side, whereas a lot of the decisions I made were business to business and on a more solid financial foundation. It took me a year to turn it around.' During that time, he formed a close working relationship with the UK office—and in particular with the new head of Celestion International, Gordon Provan.

We were dominating the guitar speaker market, to the extent that it was almost a case of asking, 'Which Celestion is in that amp?'

Provan reorganised the international subsidiaries to improve performance, and with the assistance of Wellikoff profitability was quickly stabilised across the full range of products in the United States: retail MI (musical instrument speakers), HiFi and OEM (components to original engineering manufacturers). 'We were successful in all three', related Wellikoff. 'We sold a lot of product to just about every guitar amp company. We were dominating the guitar speaker market, to the extent that it was almost a case of asking, "Which Celestion is in that amp?"'

Tighter control and a quick turnaround resulted in significant growth the following year: 'We expanded more into the bass guitar market, with small portable systems and practice amps, and we did well with the Vintage 30, Sidewinder and the G12 ... they were the most popular.'

Wellikoff worked closely with all the company's key accounts. On MI and HiFi he was able to take advantage of existing relationships with distributor/retailer operations like Guitar Center and St. Louis Music (SLM), but where manufacturers were concerned, he needed to create new partnerships. And, to establish those, he believed it was vital to meet personally with every OEM account around twice a year.

Clearly, assistance was required, and Brian Coviello arrived at Celestion towards the end of 1989; by April 1990 he had been named national sales manager for Celestion's MI and pro audio divisions.

It was agreed that business in the US had to go over and above the guitar speaker market. Although the budget miracle that was the Celestion 3 had caused a stir on its launch, the US HiFi market continued to be difficult to break into, much more so than the European territories. However, the SR Series of finished public address cabinets had potential: despite having a reputation for selling raw frame drivers, Celestion didn't sell to manufacturers in the US public address market at that time, so there was no potential conflict of interest. According to Wellikoff, there was technology on the SR units that had garnered some interest, so could systems be the key?

Coviello, who was initially tasked with guitar speaker retail sales, while running OEM with companies including Fender and Mesa Boogie, agreed. He noted there was little activity with the public address speakers: 'We were not selling enough. But because I had an extensive background

Left: The SR Series had potential for sales in the United States.

Right: Selling the classic British rock sound.

in live sound it seemed like the SR1 was a great start and this was one of the first systems with a dedicated controller. We started pushing the cabinets to our Independent Representative Network. They in turn demonstrated them to PA dealers that sold to end users and they began to sell.'

His mission was boosted still further when Yes drummer Bill Bruford, along with Rick Wakeman, started using them: 'The dealers were happier when we had more product and with PA systems, I had something I could sell apart from guitar speakers.'

The approach clearly worked, and SR systems found their way into nightclubs, venues and houses of worship across the United States. It was enough to convince Coviello of the validity of finished sound reinforcement systems as a business model such that he subsequently played a role in the design specification of the later CR Series.

With the 1990s now in full sway, the modern identity of the company was becoming clear: forged in Britain and nurtured and grown overseas, no more so than in the United States, where the company had finally begun to take root. Celestion was now seen as an innovative and international brand, both consumer and business-to-business focused; recognised for music production and *re*production. This was the era that brought the modern Celestion into being.

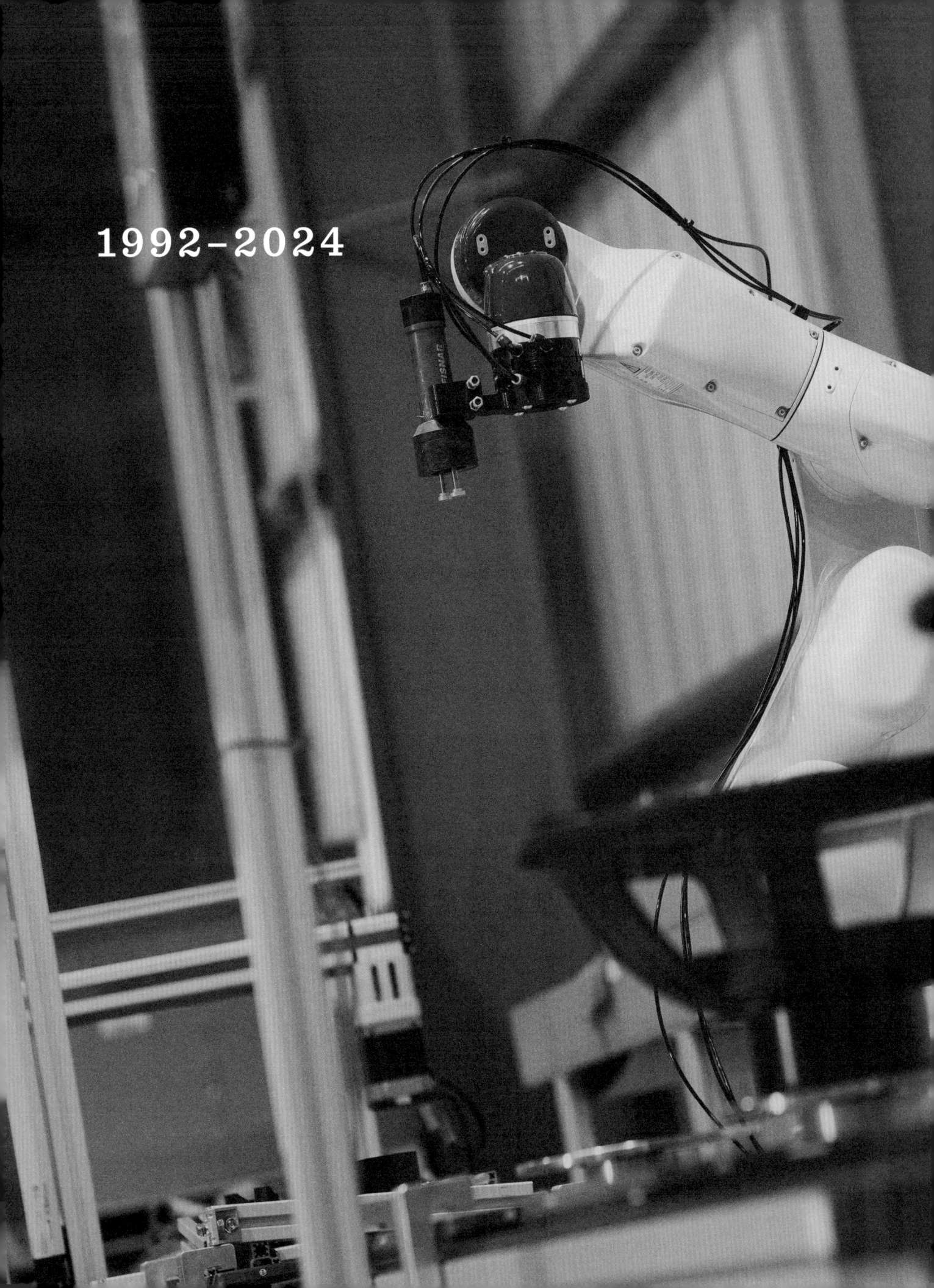
1992–2024

4
Technology Leads the Way

The robotically assisted production line undergoes commissioning in the Claydon factory.

15 The Creation of KH Manufacturing

In mid-1992, Celestion Industries sold off its audio division (Celestion International Ltd), leaving the remainder of the corporation free to focus solely on its textile business.

The organisation taking over Celestion audio was a private investment company created specifically for the buyout: Kinergetics Holdings (UK) Ltd. This consortium was formed by Gold Peak Venture Capital of Hong Kong, the investment arm of OEM technology giant Gold Peak Industries (whose stake was 50%), US audio company Kinergetics Research (40% share) and British investment company, PL Banner & Associates (the remaining 10%). Its valuation was £4.7m, including £1.6m of debt and £500,000 of goodwill, as reported in London's *Financial Times* of 2 June 1992.

At the same time, the new holding company brought into the fold another eminent name in British HiFi, Maidstone-based KEF, which at the time had been in financial trouble. It was far from a speculative play: Kinergetics Research was a credible audio company dating back to 1951. Gold Peak itself had significant roots, having begun in 1964 as a producer of batteries and subsequently grown to become a multinational force in electronics and technology, and whose chairman and chief executive, Victor Lo, had a passion for all things HiFi. Having been listed on the Stock Exchange of Hong Kong since 1984, Gold Peak had long been committed to steady growth through assiduous investment alongside skilful management and a global outlook.

While the Gold Peak Electronics Division (GPE) was already well established as a manufacturer of car audio, the Kinergetics shareholding enabled the company for the first time to position itself in the domestic HiFi and professional audio systems markets. For Gold Peak this was a significant and strategic business decision made for the long term: an opportunity to combine its expertise in marketing, product development and manufacturing with Celestion's undeniable skills in speaker design and production.

Opposite: Two Celestion HiFi designs from the 90s, the 100 and the Kingston.

Celestion Industries had declared a pre-tax loss of £1.06m to December 1991, with sales falling by 12% to £36.6m, so the time had seemed right for a sale. The existing premises was secure—as Celestion Industries had granted a 10-year commercial lease on its Ditton Works in Ipswich to the audio division—and the management team, under the leadership of Gordon Provan, Celestion chairman and managing director, would remain.

The takeover itself took place when the company was on factory shutdown during the summer. Peter Wellikoff, who by then had been running the North American operation for more than four years, vividly recalls the takeover: 'It was an interesting acquisition which led to an awkward transition, as there was a sudden cultural shift.' The deal was completed right in the middle of an important trade expo—the Summer Consumer Electronics Show in Chicago. 'So Gordon Provan and I made an impromptu press announcement.

Two HiFi Giants

An early consideration was how best to maximise the returns of these two HiFi giants through the potential synergies on offer. After all, with KEF now financially secure, one commentator estimated that the combined might of the two brands would make them the fourth largest specialist speaker firm in the world.

Initially it was confirmed that KEF and Celestion would remain independent and operate their own separate UK-based research and development (R&D) and manufacturing activities, albeit with 'more aggressive' and more 'market driven' product development programmes along with increased advertising and promotion of both brands. The board of Kinergetics (according to head of corporate development, Stuart Harris) considered the products of the two brands to be complementary, and with fresh capital injection the two manufacturers, in combination with the technological synergy brought by minority shareholder Kinergetics Research, would be expected to expand market share.

Building on the runaway success of the Celestion 3 and the rest of the unit range, Celestion pressed ahead with an upgraded version, and in 1995 the entry-level bookshelf speaker in the range, Celestion 3 MkII, was awarded 'European Loudspeaker of the Year'. Other notable models that came into production at that time included the 100 and the 300, Celestion's only speaker of transmission line design. There was clearly no sign of a let-up in product development.

The previous year, which was also the company's 70th anniversary, saw the launch of the impressive and beautiful Kingston to much fanfare. As part of the launch publicity, Managing Director Provan had announced to the press that Celestion had doubled in size since 1988, achieved primarily due to the success of the Celestion 3 and its siblings. It was no mean feat during a time period that saw all the uncertainty that comes with a corporate takeover played out against a backdrop of extreme financial uncertainty. A lengthy recession had haunted

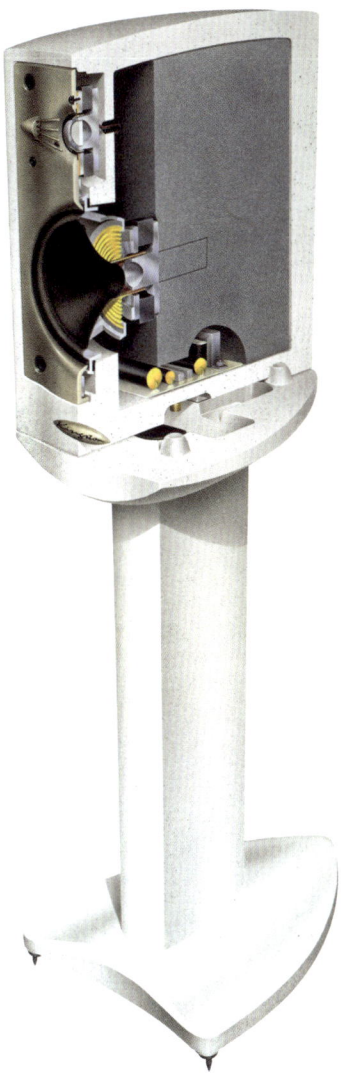
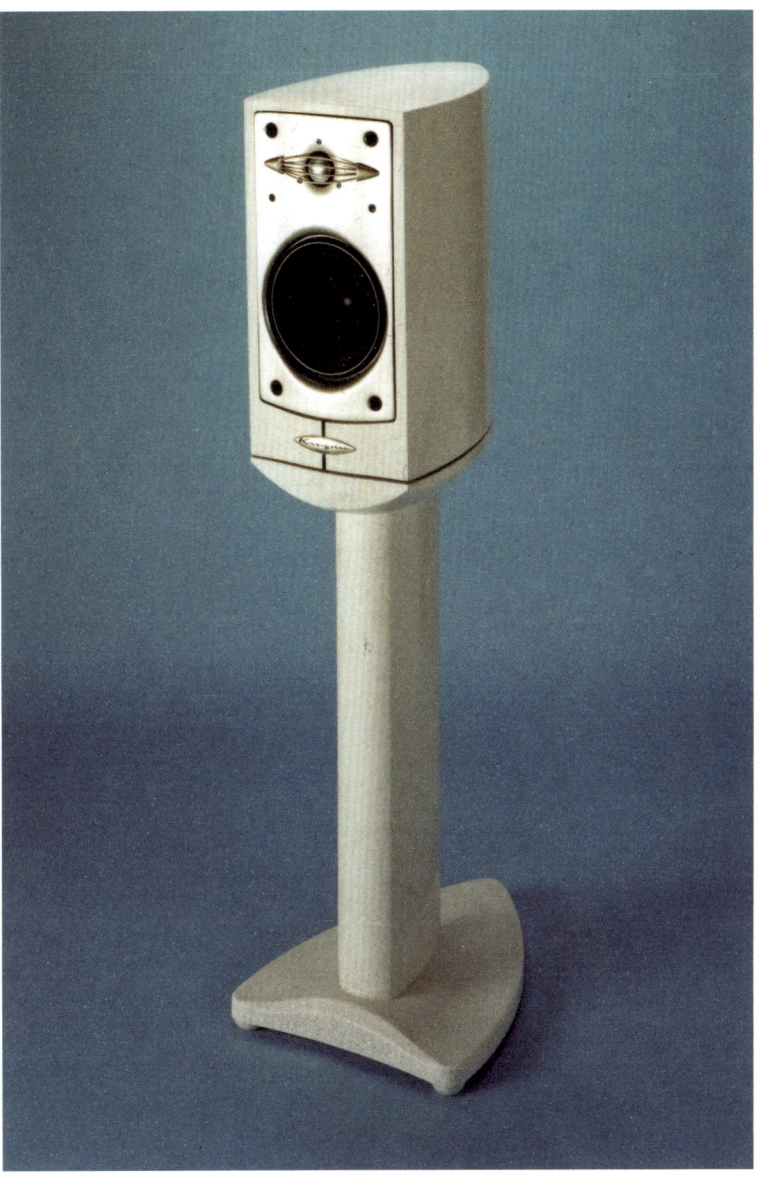

Left: Artist's concept of the Kingston.

Right: The Kingston itself.

Britain since the beginning of that decade, which had been compounded by a currency crisis that was precipitated by the UK's hasty withdrawal from the European Exchange Rate Mechanism, an event later referred to as 'Black Wednesday'.

The Kingston was the embodiment of a new-found confidence. Intended as the ultimate incarnation of the ground-breaking SL6 and its family that had been introduced more than a decade previously, the Kingston incorporated many of the landmark technologies of its forbears, housed in a unique 'Alpha Crystal' enclosure.

The Alpha Crystal was a mixture of mineral and a high-density, inert polymer, with a high internal loss (so very low resonance) and with the ability to be sculpted relatively easily. The Kingston was moulded into a unique shape that was capable of suppressing standing waves and avoiding sound wave diffraction.

Left: Home Theatre Systems advertising.

Right: The A Compact bookshelf speaker.

However good it might have sounded, the Kingston was a commercial disappointment. The high price of the product, coupled with concerns about the durability of the Alpha Crystal material, led to lacklustre sales. Additionally, with 'home theatre' products and the all-in-one packages starting to populate the houses of the general public in the 90s, aided in no small way by the arrival of the Dolby Digital® surround circuitry, the market seemed to be heading in a different direction.

The subsequent A Series was built with those innovations firmly in mind. Now adopted as Celestion's premier range, it was marketed as combining 'exceptional sensitivity with high power handling capabilities to give an unusually broad dynamic range'. All cabinets were magnetically shielded, making them well suited to home theatre applications. The cabinets were of very conventional material and design and, being very heavy duty, were solid performers and sonically more than competent. The A1 and A Compact in particular were able to tease impressively big sounds from small boxes.

June 1996 saw an announcement from then managing director, Andrew Osmond, that Celestion would be split into two parts: the Professional Division (musical instrument speakers, transducers for professional audio and finished sound reinforcement systems) and the Consumer Division (HiFi and home theatre), with KEF's managing director at that time, Ray Lepper, appointed managing director of the latter.

Following this formalised separation, Celestion HiFi was duly relocated to Maidstone. The KEF headquarters also became the centralised hub for all administrative functions for the next decade and a half.

Soon after, Celestion and KEF brand manufacturing was fully united under the banner of Kinergetics' KH Manufacturing Ltd, which became the single corporate identity in which both KEF and Celestion resided. Some of the Celestion engineering team, including Bob Smith, had been seconded to the KEF headquarters (where the A Series range had finally been completed), and would regularly travel to Kent to supervise.

Celestion HiFi development continued, with 1998 bringing the attractive C Series, using the latest computer-aided design techniques, and constructed from extruded aluminium. Also added were more 'lifestyle-orientated' products, designed to have a broader consumer appeal, such as the compact and durable MP1. Sadly as a new millennium dawned, the Celestion HiFi line slowly seemed to run out of impetus.

By the time Frank DiGirolamo took over executive responsibility for the Professional Division in 2000 (having already become *de facto* managing director of KEF as well as CEO of the Group) one of his main missions was to consolidate HiFi into a single brand for the new millennium. With KEF appearing to be the dominant brand, it became increasingly clear that the writing was on the wall for Celestion HiFi.

A Switch in Retail Strategy

Despite assurances to the contrary, Celestion HiFi sales in the United States were an early casualty of the change in ownership, finally resulting in the US office in Holliston, MA, being closed down. The OEM (components to original engineering manufacturers) chassis speaker sales brief was being handled remotely by the UK sales team, in particular by the recently appointed Paul Airey. When it came to retail sales, the decision was taken to completely outsource. In 1996, established audio distributor Group One, headed by Jack Kelly, was appointed specifically to handle sound reinforcement systems and retail transducer sales; it was an arrangement that existed for the next 14 years.

Assessing the Celestion brand value, Kelly believed Group One could help with the cabinet sales and finished sound system products in particular. Yet, like his predecessors, his team were soon wrestling with the dichotomy of trying to market complete systems into an arena of high-pedigree domestic brands, when Celestion's own pedigree and heritage (on that side of the Atlantic at least) lay firmly in its guitar speakers. It was at the 1999 NAMM Show in Anaheim, CA, that the penny finally dropped. 'There was a lot of PA hung on truss and in one corner a small stack of guitar speakers. There were no visitors on our booth and yet the show itself was packed. It was a total disaster. We were just not important in that market space. I came to the

Exhibiting finished sound reinforcement systems at the NAMM show in the late 1990s.

conclusion we were competing with our target PA chassis customers while going up against companies such as JBL who had an established footprint of PA cabinets,' Kelly rationalised.

Seeing the situation, Group CEO Frank DiGirolamo turned to Jack Kelly for a solution: 'It was guitar speakers that were the holy grail. I told him the whole booth should be reversed, pushing guitar speakers to the fore.' In many ways, it was a return to Bill McGrane's ethos of the 1980s. Hence, Group One initiated a perception change—promoting Celestion drivers and moving the raw frame speaker brand from being simply a part to a product in itself.

Focusing on Technology

In the engine room that was Celestion's Foxhall Road R&D department, developments were surging forward as the adoption of new software tools brought additional impetus. In 1992 Graham Bank established the company's first mechanical Finite Element Analysis (FEA) set-up, a computerised method for predicting how a product might react to real-world forces such as vibration, which opened up immense possibilities for future acceleration and streamlining of the development process.

According to Paul Cork, Celestion's first use of mechanical FEA in the Professional Division was to resolve a quality issue it was having with a bass guitar speaker: 'The dust cap fitted on the coil former which allowed the cone to distort, consequently the coil deformed and hit the magnet assembly causing a "buzz". Fitting the cone with a bigger dust cap braced the cone, preventing it

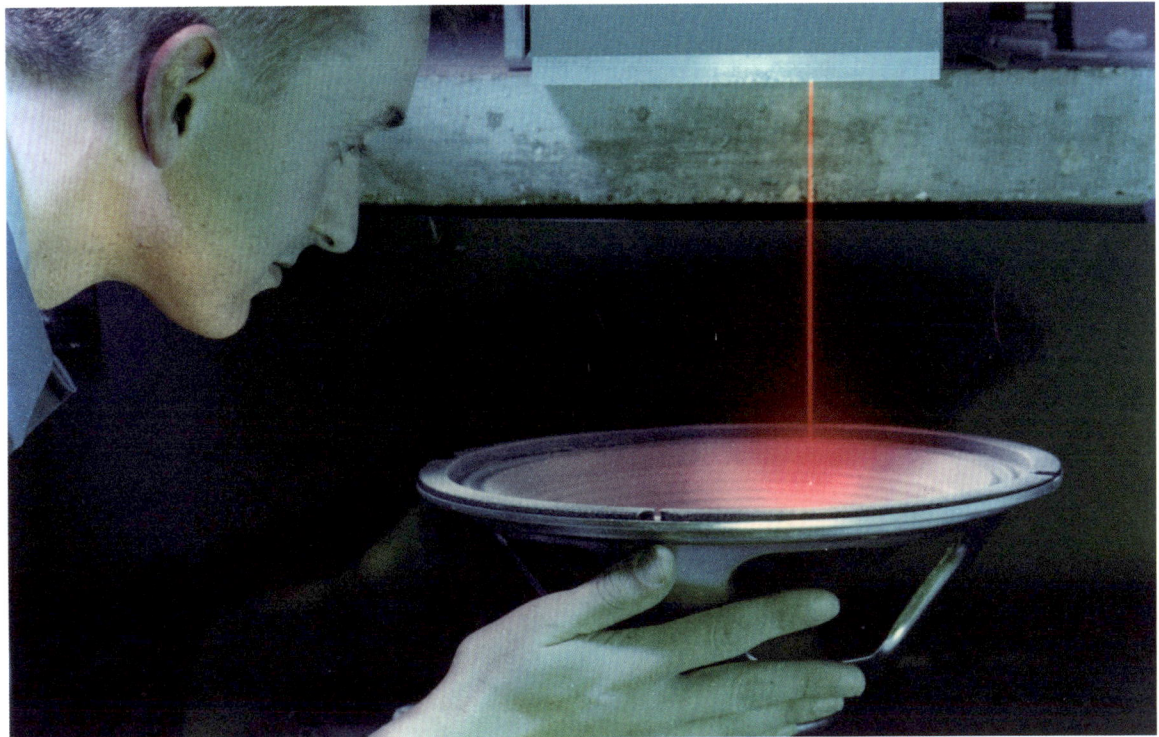

Paul Cork measures a G12 using laser interferometry.

from "waggling" as much and thereby preventing the coil from deforming and hitting the magnet assembly.' The FEA process enabled the team to rapidly identify the viable fix for the problem and swiftly bring it to production.

Bank had very much been the concept man, tasking others with converting inspirational ideas into reality. That had been the case with the Laser Doppler Interferometry system, which Gordon Hathaway had brought to fruition, and so it was with FEA, this time with Julian Wright in the hot seat, refining the process: 'FEA had just come out of university and interfaces were crude, it was like coding with script files. It was very immature and not user friendly and so I was writing interfaces to make it more accessible.' However, Wright could see the power that the FEA process offered: 'We quickly latched onto a version which enabled us to couple mechanical and acoustical behaviours at the same time. You need to be able to couple the cone to the air around it.'

After this, the process became substantially more valuable. Previously, it had been very expensive and time-consuming producing prototype after prototype of magnet assemblies, but FEA enabled the company to simulate physical phenomena relatively easily. As a result, it was now able to deliver a couple of models a day with FEA, even though it was at a fairly primitive stage. For compression drivers, in particular, FEA would be a game changer, shortening development times and improving product quality, largely because of the tight tolerances with those transducers. This was an advantage that the company could exploit, and it would do so in the coming years. Not for the first time, technology was leading the way for Celestion.

16 Celestion Professional

HiFi was the key incentive for Gold Peak's acquisition of Celestion, and it was HiFi that was the source of most of the growing pains for Celestion as the company began another new chapter, this time as part of a much larger corporation. But in many ways, it was business as usual for the new Professional Division, which encompassed 'power' (guitar speakers and small PA speakers) 'heavy power' (large-format woofers) and sound reinforcement cabinets.

Absolute Sound

After the success of the SR Series, Celestion was looking for a follow-up. The CR range of carpet-covered sound reinforcement cabinets, featuring the company's self-designed, workmanlike RTT50 tweeter and the painted cabinet version for fixed installation dubbed CRi, although excellent proving grounds for the home-designed drivers, largely went under the radar when they were released in 1993–1994. However, recognising a growing market desire for plastic injection-moulded speakers for installation, in 1995 the company produced the sleek KR Series.

Probably as a result of the burgeoning nightclub and music bar boom, a range of venues started recognising the value of good sound, from theatres to houses of worship. KR met these needs, aesthetically pleasing and designed specifically to fit into multipurpose venues requiring fixed sound reinforcement, with its compact footprint, low weight and easy installation. The range, comprising three different sizes of cabinet plus a subwoofer, ticked all boxes; moreover, Flexirol bass driver technology, deployed in the woofers of the KR8, KR10 and KR sub-cabinets, had another day in the sun, offering better low-frequency results and greater excursion control.

Meanwhile, work on developing the highly significant Road Series was beginning—a roto-moulded MI/PA solution which for the first time would take a full, affordable Celestion system into the heartland of the gigging musician. But its gestation period was long, and the millennium was at an end before it emerged in the marketplace.

The rotational moulded cabinet was unique to Road and thus the key selling point, but there were delays, as the company struggled to find the right partner to produce it. Ian White was head of

Opposite: Power speakers (guitar and pro audio), circa 1994.

The robust Road series was built for the rough and tumble of touring.

engineering at the time: 'We wanted to make a plastic cabinet that was sufficiently robust to withstand the rough and tumble of touring and were interested in rotational moulding since there was only one other company making a similar loudspeaker.'

'We had a relationship with a company … and tried for a long time to make a cabinet but it kept warping and creating all sorts of other problems with cosmetics. In the end we had to switch suppliers and then it all suddenly came together.'

White could foresee many advantages of using rotational moulding to create a solid plastic skin, and combining this with a foaming agent to add thickness. Celestion also applied for a patent for stiffening tubes.

The material itself was robust. 'In reality the construction was almost like a composite providing the stiffness and the resilience to it getting knocked and banged. Also, the tooling was not very expensive, so it had a lot going for it,' said White: 'Another advantage which we stumbled on was that we moulded the horn flares into the plastic, so it didn't need to be made in two halves like injection moulding but was a complete entity. That gave us the advantage of being able to insert the compression driver to abut the horn at the rear. Having that structure go all the way through provided an extra stiffness and was quite a neat application. It was certainly an interesting engineering project.'

We wanted to make a plastic cabinet that was sufficiently robust to withstand the rough and tumble of touring

As something of a 'range filler' during the delay in bringing Road to production, Celestion released the CX range, using all the Road acoustic work but presenting it in a painted plywood cabinet. This was followed by the cost-effective QX range in 1998—essentially a cost-reduced revamp of the CR carpet-covered cabs. By using four-layer coils on the low frequency (LF) to get enough inductive rise to effectively cut off the response at around 3kHz, the cost of a separate crossover was avoided.

By the turn of the millennium, the QXa Series—a QX with an amp module inside—was produced. Fairly sophisticated for the time, this active system was designed and made by GPE for world markets, featuring a built-in Class D amplifier.

Then came the premium installation solution, CXi (built with solid 18mm birch plywood and available in both black and white), which also provided a magnificent curtain raiser for the new millennium. Made with OmniMount-compatible hardware, several models featured the new CDX1-1750 1in exit compression driver (the first of what would later evolve into Celestion's CDX Series of high-frequency (HF) drivers, developed by Mark Dodd). Marketed as *Absolute Sound*, the CXi Series found itself installed in venues globally and carried the banner for Celestion sound reinforcement systems into the new century.

The New, the New Old and the Customised

As Celestion's sound reinforcement systems made their mark, over in the 'power' department, the desire for 'the old sound', when it came to guitar speakers, continued unabated, and the groundwork laid by the development of the Marshall G12 Vintage (Vintage 30) continued to yield dividends. The Vox Blue, now firmly considered to be the 'holy grail' of guitar tone, was to live again, this time labelled Celestion Blue.

To resurrect the classic alnico speaker, the magnet assembly was recreated identically, with the same grade of alnico and the voice coil former wound to the exact same specification. Just as it was with the Vintage 30, in order to hit the tonal target, the key focus was the cone. Again, the original was no longer available, so a new cone was tooled and, though not identical, was sonically close enough to the original to win approval. So it was that the Celestion Blue earned its own individual T numbers: T4427 for the 8Ω unit and T4436 for the 15Ω unit (instead of 16Ω, a nod to the vintage specification).

Newly relaunched by 1994, the Celestion Blue (*left*) and the G12M Greenback (*right*).

It was the use of contemporary materials in the cone manufacturing process that likely prevented the new cone from being 100% identical to the original, but it was the application of science—Celestion's trusty Laser Interferometry—along with a significant amount of listening, that enabled the original to be so closely mimicked.

Legend has it that amp guru and noted tone chaser, the late Ken Fischer of Trainwreck Circuits fame, was also consulted on the voicing of this illustrious reissue, perhaps adding a further patina of authenticity, although little corroborating evidence can be found.

The Celestion Blue was to become the flagship of the company's 'Specialist' guitar speakers range, accompanied by a new reissue of the G12M 'Greenback', complete with green rear can and rebuilt to the exact original specification, as the rear label proudly stated. This was joined a couple of years later by a reproduction of the G12H rebadged as the 'Anniversary' in time for the company's 70th birthday celebration in 1994. Just like the original, it was a heavy magnet version of the G12M but, confusingly, built without a green can—perhaps as an homage to the T1134, the earliest ceramic magnet G12 of all.

A growing range of specialist guitar speakers was marketed under the strapline 'The Voice of Rock & Roll'.

Also in the range were the, by now, well-established Vintage 30, the Classic Lead 80 (a rebadged G12-80, from the late 1970s range), an S12-150 (formerly the Sidewinder), a newly revamped G12L, as well as the short-lived and often maligned Modern Lead 70, plus the Vintage 10 and Vintage 8, both designed in 1991 to capitalise on the success of the Vintage 30.

A further range simply categorised as 'Guitar Loudspeakers' consisted of speakers that bore the standard Celestion nomenclature: upper case 'G' followed by the speaker's diameter, then a letter that denoted the magnet size and weight, and finally the power rating. Of the range available in 1994, the G10L-35, G12K-85 and G12T-75 proved the most enduring.

Up until this point, the 'G' stood for General Purpose and the G Series speakers were used for all sorts of sound reinforcement applications. Then, all of a sudden, 'G' was simply for 'Guitar', the speakers built with a paper edge. New ranges with new classifications, purpose-designed for other instruments (bass and keyboards) and for sound reinforcement, built with a cloth edge, began to appear.

Celestion continued to supply a wide range of amplifier brands, not just with catalogue product but modified units too, ranging from slight alterations to standard units to fully customised speakers.

Chief among these brands was Marshall, which remained a principal customer for all things guitar speaker. It had kept the Celestion R&D team busy throughout the 1980s with the likes of the aforementioned G12 Vintage and the G12T-75, and into the 1990s there was a steady stream of custom requirements, notable models including the G12B-150, the Marshall Heritage and the V12-80, better known as the 'Wolverine'.

Current Head Engineer Paul Cork remembers his first guitar speaker project as a rookie technician: 'The first speaker with my name on it was the G12T Special, T4342 in early 1992. It was the first speaker for the Marshall Valvestate range at the time when Marshall first decided to branch into hybrid valve/solid state amplification.'

'As technicians we had to prep demonstration cabinets. Steve Grindrod, Chief Engineer from Marshall came into the lab one day for a critical listening test and as I was fitting the speaker into the cab the screwdriver slipped and went through the edge of the cone, rendering the speaker useless. Everyone in the lab has done this at one point or another—but it was very unfortunate timing for me, and very embarrassing!'

And it wasn't just lead guitar speakers that were going to Marshall, but bass guitar units too. Though, unlike those produced in the early 1970s, modern bass speakers were built like PA speakers: more robust, with greater power-handling capability and more excursion, better able to cope with reproducing the bass guitar frequency range.

As testament to the long-enduring relationship between Celestion and Marshall, on the occasion of Celestion's 70th Anniversary Jim Marshall gifted Celestion one of the commemorative full stacks the company had recently produced for his company's own 30th birthday. Wrapped in a beautiful blue vinyl, it featured a monster three-channel 100W amp sat on two G12T-75-loaded, 4x12 cabinets (one straight, one slanted). An inscription plate on the reverse reads:

> Congratulations on your
> 70th ANNIVERSARY
> from MARSHALL AMPLIFICATION
> Manufacturers of the Largest Volume of
> CELESTION LOADED
> GUITAR AMPLIFICATION
> in the World
> 1994

Left: Jim Marshall presents Gordon Provan with a commemorative Marshall stack in celebration of Celestion's 70th anniversary.

Right: The Marshall stack, in blue vinyl, was loaded with G12T-75s and inscribed on the rear.

While the biggest customer, Marshall was far from the only amp company: Vox had been a regular consumer of alnico speakers and Greenbacks plus a variety of other models. Orange took a customised version of the G12T-75 for a time, as well as standard heavy magnet products such as the G12H, Vintage 30 and G12K-100, and Laney had taken a wide variety of standard 12in and 10in products to suit its comprehensive range. Around this time, German manufacturer Hughes & Kettner, part of the Stamer company, began a long-term partnership, culminating in the development of a number of custom guitar speakers under the 'Rockdriver' label.

It was around the end of the 1990s that Celestion began experimenting with neodymium, a relatively new magnet material that was increasingly being looked at for use with transducers

Trace Elliot was a big UK name in bass amplification for which custom bass units were made. They had also tried their hands at lead guitar amps, at one time requiring a 5.3Ω variant of the Vintage 30: a strange impedance in the world of guitar speakers that could only have been conceived of for use in a 3x12 cab.

Across the Atlantic, Fender—then as now—was the dominant amplifier brand as well as Celestion's biggest US-based guitar speaker customer. It was on a trip to the Fender head office alongside US OEM Sales Manager Paul Airey, to successfully head off a potential crisis with some faulty 10in speakers, that Ian White conceived the G12T-100 (aka Hot 100). With its 2in voice coil, neutral voicing and massive headroom, this speaker became Fender's go-to speaker for the 1x12 Prosonic and later for its first digital modelling amp, the Cyber Twin.

Mesa/Boogie was another US amp company that had turned to Celestion due to its desire to add a 'British' sound to its portfolio. First, it adopted a customised version of the G12-80 (which became the Mesa C90) in the late 80s, and by 1991 had also gravitated towards the Vintage 30, as many were doing at that time.

Elsewhere stateside, St. Louis Music, the parent company of Crate, had had custom speakers built with a bright green chassis (the V8, V10 and V12 designed by Ian White), and bass brand Ampeg had loaded Celestion bass speakers. Ampeg also produced lead guitar amps, and for a time these were loaded with the company's own-branded version of the G12L. Another notable custom unit designed by Ian White was a 32Ω, 8in bass driver for SWR, the BG8-60, which was deployed in the now-legendary Henry the 8x8 cabinet.

It was around the end of the 1990s that Celestion began experimenting with neodymium, a relatively new magnet material that was increasingly being looked at for use with transducers. Neodymium has much greater flux density than the conventional iron-based ceramic magnets that were and still are most commonly used for loudspeakers. A smaller magnet requires less metalwork around it to create a magnetic circuit, and as a speaker's magnet assembly is a large

The G12 Century, the world's first neodymium magnet guitar speaker.

contributor to the overall weight of the transducer, neo magnet speakers tend to be much lighter. Neodymium also gives the designer the opportunity to increase the overall strength of the magnet for tighter control over the voice coil and more output.

One of the drawbacks of neodymium is that it can lose some of its magnetic field strength when exposed to high temperatures, resulting in variable performance over time. One solution is to create additional means of 'sinking' heat, conducting it away from the magnet assembly by increasing the surface area of the associated metalwork, say by designing the assembly with fins.

The emergent FEA technology enabled Celestion R&D to create a light but powerful neodymium magnet assembly for a new guitar speaker, with a striking star-shaped aluminium heat sink design to take care of the cooling issue. In stark contrast to the beginning of the decade, which was all about attaining the vintage sound, the resultant G12 Century was the very model of a modern-looking and modern-sounding guitar speaker: loud, clean and accurate.

The BX Series of Precision Sound Reinforcement Loudspeakers.

Pro Audio: Technology Inverted?

Just as front-of-house systems were becoming more sophisticated and amplifiers notably heftier, the design of sound reinforcement transducers were becoming more involved. PA was now more than simply adding a cloth edge to a general-purpose loudspeaker and perhaps giving it a more heat resistant voice coil and an aluminium chassis.

If the B15 and B18 woofers, produced in the second half of the 1980s, stated an intention to build serious PA drivers for serious sound reinforcement, then the BX Series was the next generation in that regard. Marketed as 'Precision Sound Reinforcement Loudspeakers' they boasted a host of advanced features 'to extract every last ounce of performance', while the unique three-legged chassis and distinctive rubber moulded magnet cover gave the range a no-expense-spared feel of high quality.

Taking what they'd learned from developing this new generation of low-frequency loudspeakers and the expensive, bespoke chassis, Celestion also developed a parallel range of car audio speakers, a new departure for the company, in a hotly contested market niche. Celestion 'Autosound' drivers were built to reproduce 'exceptionally accurate bass' in small vented or infinite baffle (completely sealed) enclosures.

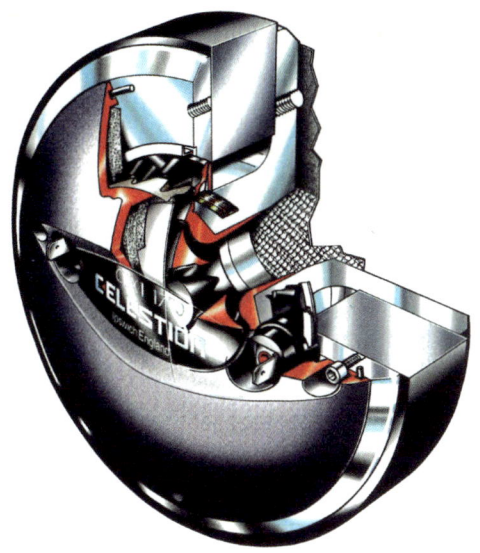

The CDX1-1750 HF compression driver.

The speakers formed a coherent range and were marketed as providing a musical, satisfying low end, with high efficiency, low distortion and incredible power handling. More proof, if it were needed, that Celestion was still much more than a HiFi company and was right at home designing and building loudspeaker drivers regardless of application.

New generations of low-frequency speakers continued to be developed as the 'power' line diversified. The utilitarian G Series, now restricted to being guitar speakers, were superseded first by the K Series and then by the Q Series, both cloth-edge-type woofers whose role was to produce the kind of sturdy low end demanded by cost-effective fixed-install cabinets and keyboard amps, with the BG Series dedicated specifically to bass guitar cabs.

By the end of the decade, the best that Q and K had to offer began to be manufactured at GPE after being rebranded as Truvox, in homage to the company's PA heritage.

Retail guitar speaker sales were ever buoyant in the United States under the supervision of Group One, and it was due to the desire to ride in the slipstream of their success that the retail-focused Frontline cast frame woofers were created. Intentionally targeted at the sound system community based around Canal Street in New York City, they offered slightly more budget-friendly bass than the BX drivers.

It was the need for HF units to deploy in Celestion finished systems that first initiated the in-house development of a new HF compression driver. During the early 1990s, compression drivers had been bought in from other transducer companies to load into finished cabinets, a practical necessity at the time, but not optimal for a company that was awakening its desire to build pro audio transducers in-house.

CELESTION PROFESSIONAL

The 600W NTi-1550, with front-mounted magnet assembly.

Earlier compression horn driver designs, brought into being by Les Ward and his contemporaries, had been intended for speech, warning or announcement systems and the like, or, in the case of one particular design, a foghorn. What was required was a design that was based on a principle already successfully used in sound reinforcement for live music, and this was how the CDX1-1750 came into being.

Much of this work and the ensuing compression driver programme was masterminded by Mark Dodd. Transferring to Celestion from KEF in the mid-1990s, he had previously been chief engineer at Tannoy, where he had worked on coaxial drivers and was instrumental in the development of the 'tulip waveguide'.

As FEA was becoming a more established tool at Celestion, with its high-powered computational capabilities, Dodd was soon probing heavily into the simulation possibilities that the technology provided. This would leave him perfectly placed to lead Celestion's drive forward into the development of pro audio HF, which would ultimately pay significant dividends.

'I learned how to use finite element methods in anger on designs,' he said, adding that FEA was particularly suitable for high-frequency drivers. 'As a result, the time taken to set up a [computerised] model was cut by a factor of at least 10 so that the virtual prototypes could be directly compared to measurements made in the lab. I believe this was a first.'

The deployment of a neodymium magnet with a 600W woofer was another industry first and brought Celestion pro audio firmly into the 21st century. This speaker had a much more unconventional solution to the task of heat dissipation than conventional woofers or the G12

> The product was awarded a patent for its novel topography of magnet and steel, while exhibiting superb sonic detail at an incredibly low weight

Century neodymium guitar speaker. Rather than a heat sink, it instead had the entire magnet assembly mounted at the front of the driver, enabling the heat to freely dissipate into the air unencumbered by any cabinet, and was known as NTi: Neodymium Technology *inverted*.

The product was awarded a patent for its novel topography of magnet and steel, while exhibiting superb sonic detail at an incredibly low weight, and, according to Paul Cork, 'made everyone think more about heat restraints on neodymium'.

Manufacturing Moves East
The eventual shift of much high-volume product manufacture from the Ipswich factory to the GPE manufacturing base in Huizhou, Guangdong Province, China, towards the end of the 1990s was viewed with mixed feelings. At the time there was a growing fear of Western intellectual property being plagiarised by Chinese companies, enabling copycat products to flood the market. But no such fate awaited Celestion, which was now part of a respectable and highly profitable global organisation with an established trade base. Having such access also gave the company a real competitive edge.

Ian White summed up the prevailing mood: 'The takeover by Gold Peak was positive because it gave us the opportunity to source components ahead of everyone else in China for many years, while our competitors were [still learning and] making mistakes. This gave us a huge advantage. The Chinese company was good at manufacturing, and it would approve a supplier in a strategic way so that all the companies they put us in touch with were good.' Not only that, it was clear that Gold Peak was also willing to invest in the process to make it successful in the long term.

As a policy, it tended to be higher-volume, lower-cost products that were transferred to GPE. The benefit was that higher volumes of product could be produced faster at lower cost—an advantage that made business sense to ensure the continuing success of the company.

Transfer of manufacture reached critical mass after the turn of the millennium, as it was decided to move premium guitar speakers across to GPE, a move that finally sealed the fate of the aged and increasingly shabby Ditton Works factory, which was deemed too large for the diminishing workforce and a little too decrepit to fix up. Celestion's tenure on Foxhall Road was finally over—it was time to move on.

17 New Beginnings

August 2000 saw the demise of US-based Kinergetics Research, President Tony DiChiro attributing the company's shutdown to 'a shrinking market for high-quality two-channel audio'. Shortly before, Gold Peak had taken the opportunity to acquire the remining shares in Celestion's parent company, and by the middle of 2001 Kinergetics Holdings had metamorphosed into GP Acoustics.

By now, Celestion had transferred nearly all of its raw frame speaker manufacturing to Guangdong Province, with the remaining HiFi and systems cabinets manufactured either in China or at the KEF plant in Maidstone. The logical next step was to move the remaining Celestion staff to a smaller facility, one with lower overheads and that didn't require quite so much maintenance as the venerable Foxhall Road.

A site was chosen at Claydon Business Park in Great Blakenham, a commuter village just north of Ipswich, and a brand new building was commissioned. The uprooting from Foxhall Road had initially caused a major upheaval, not least because most of the staff lived close to the old factory, east of the town, and the new plant was to the north-west. The change had come as a surprise to many on the shop floor, and only a small number opted to make the move, while many more were let go in the months leading up to the transfer, as onsite manufacturing dwindled.

A Fresh Start

Right from the outset, it was clear that the new plant would be an engineering-led establishment. With the bulk of manufacturing now taking place elsewhere, the focus would be 'designed in the UK'.

In consultation with his R&D team, Ian White ensured that Claydon, for all its scaled-down size, would boast a state-of-the-art infrastructure, resulting in some great improvements over the tired, 'old industry' feel of Foxhall Road. This included a new laboratory complete with 2-pi hemi-anechoic chamber and a room purpose-built for analytically listening to new products.

Opposite: The hemi-anechoic chamber at the Claydon factory.

For the first time, Celestion engineers would have a listening environment that matched the quality of their other research tools

Reflecting on the anechoic chamber, Paul Cork recalled the problems that had been endemic at Foxhall Road: 'Back in the day we would press our own steel chassis. We had a 300-ton press for the bigger speakers, but the cycle of the press messed with the measurements in the anechoic chamber, so you always had to time your measurements accordingly, to avoid any ... interference.'

The listening room in particular proved a valuable addition, as the only listening environment available at Foxhall Road had been a redeployed meeting room, and any time speakers were tested there, the windows would rattle in the entire building.

The Claydon listening room was intended to be an entirely different experience. Built by renowned acoustic designer, Philip Newell, the new space was created along the lines of a recording studio: acoustically neutral with bass traps in the ceiling and three of the walls, the fourth wall uneven to act as a diffuser. It was a room within a room, adding acoustic insulation all around. Most vital of all was the floating floor, which was necessary to isolate the room from vibrations caused by the well-used Norwich to London railway line, which ran just 150m to the north of the new factory.

For the first time, Celestion engineers would have a listening environment that matched the quality of their other research tools, enabling them to evaluate and appraise prototypes of the speakers they were developing.

Nevertheless, the move had been a personal wrench for old hands, like Cork: 'It breaks my heart to think what we left behind at Foxhall Road—for instance a box of NIB cones for Marshall, which I thought I'd never need again.' These, in fact, were the NIB-stamped cone variant, which was standard on the CT7442 10in speaker model used by Marshall from the late 1960s to the mid-70s, now something of a collector's item.

While the new plant was being constructed, Celestion engineers were concurrently exporting the manufacture of the company's premium guitar speakers, those built on the so-called Forth line, out to the GPE factory in China. 'It was a challenging time, to get them produced over there whilst absolutely retaining the tone,' reflected White. The carefully managed process was successful: the GPE workforce adapted readily, and ultimately the migration happened with remarkably few hiccups.

The Philip Newell-designed listening room.

The original concept for Claydon had been for no manufacture to take place at all, with the only speakers built being those intended as samples to establish new concepts for OEM builds. Custom speakers for specific manufacturers now being the 'lion's share' of the business, this was an important function in the design cycle.

At the time, many key customers had been moving their product manufacture to offshore contract manufacturers in the Far East, so readily accepted the move to China-built speakers, fully understanding the business benefits. However, one customer in particular refused to accept anything but UK product. Only after a visit to that customer with GP Acoustics CEO Frank DiGirolamo in tow, was OEM Sales Manager Andy Farrow able to convince the chief executive that the Claydon plant still needed a full production line.

By the time the Claydon factory was finally open for business, Celestion had installed a small production facility to build guitar speakers, allowing the retention of some of the skilled production workers who had agreed to commute to the 'other' side of Ipswich. It was a decision that was beneficial straight away and was to prove pivotal for the company in later years.

Celestion Blue guitar speakers await finishing on the Claydon production line.

A Focused Strategy

For much of the company's history, the United States had been an itch that couldn't quite be scratched. Many successes had been achieved, but not consistently over a sustained period of time.

Through the late 1990s and into the 2000s, following the closure of the Massachusetts office, OEM sales to US manufacturing brands were taken care of by a British sales manager travelling stateside every few weeks to visit key customers, check the temperature of the relationship and unlock any new business that was to be had. Paul Richardson had taken on that role at the turn of the millennium, continuing to expand the portfolio of US clients.

Group One was firmly embedded and was taking care of the retail distribution business, but it was business-to-business sales more than retail sales that meant growth, and DiGirolamo concluded that what was required once more was boots on the ground. In the end it was Farrow, who had only recently returned for a second stint as a Celestion employee, who moved to the US with his young family. Richardson remained in the UK as Head of Sales and Marketing with the brief to build a new team for a new era.

As Head of the GP Acoustics (UK) Group, Frank DiGirolamo was focused mainly on the consumer-led HiFi and home theatre products which saw a greater financial turnover. This left Celestion Professional in need of closer strategic management, someone to breathe new life into a strong brand that was currently underperforming.

> By focusing on transducers we could design much better products and speed up their time to market

That person turned out to be Nigel Wood, and on his arrival in late 2004, the one thing that was immediately obvious to him, with the sidelining of HiFi and the slow but steady increase in the sale of raw frame transducers to the professional audio market, was that Celestion's product range had dissipated and become heavily diluted. Foremost in his mind was that by producing complete systems as well as raw frame transducers, Celestion was in effect cannibalising its own market.

Six months later he was further able to take an industry view when attending Prolight+Sound, the annual industry expo in Frankfurt, where he could see 'many people making boxes but not many making drivers'. For Wood, it was a 'lightbulb moment'.

The more he looked at the future potential of the company, the further sound reinforcement systems disappeared in the rear-view mirror. 'It simply had no future,' he recognised. By focusing on transducers, Wood decided, 'we could design much better products and speed up their time to market'. At the same time, he also believed that Celestion needed a greater balance of sales personnel to technicians, and a lot of investment generally across the board.

While he was confident that guitar speaker activity was well managed and would continue to grow healthily as a business, pro audio transducer development was severely undernourished: 'We had a limited range, so we needed to rapidly develop this. We employed some engineers and also further developed our existing software tools to assist.'

Under the direction of Wood, an almost entirely new team was built. Over in the United States, in a departure from the recent past, Farrow began to develop a regionally based, US OEM sales department. Recruits included guitarist Rick Skillman, who initially undertook a guitar speaker sales role, and, later, experienced loudspeaker sales manager Ralph Nichols, who had previously worked a stint at pro audio transducer competitor B&C, so was well qualified to seek out potential business in this area.

Sales coverage in Europe was also expanded, and 2006 saw the recruitment of China-based Kevin Yeung, who was given his own mission to build a team selling to the many Chinese contract manufacturers. An additional, much tougher target was to be the growing market of domestic audio equipment builders, and, being based in Guangzhou, he was perfectly placed to do that.

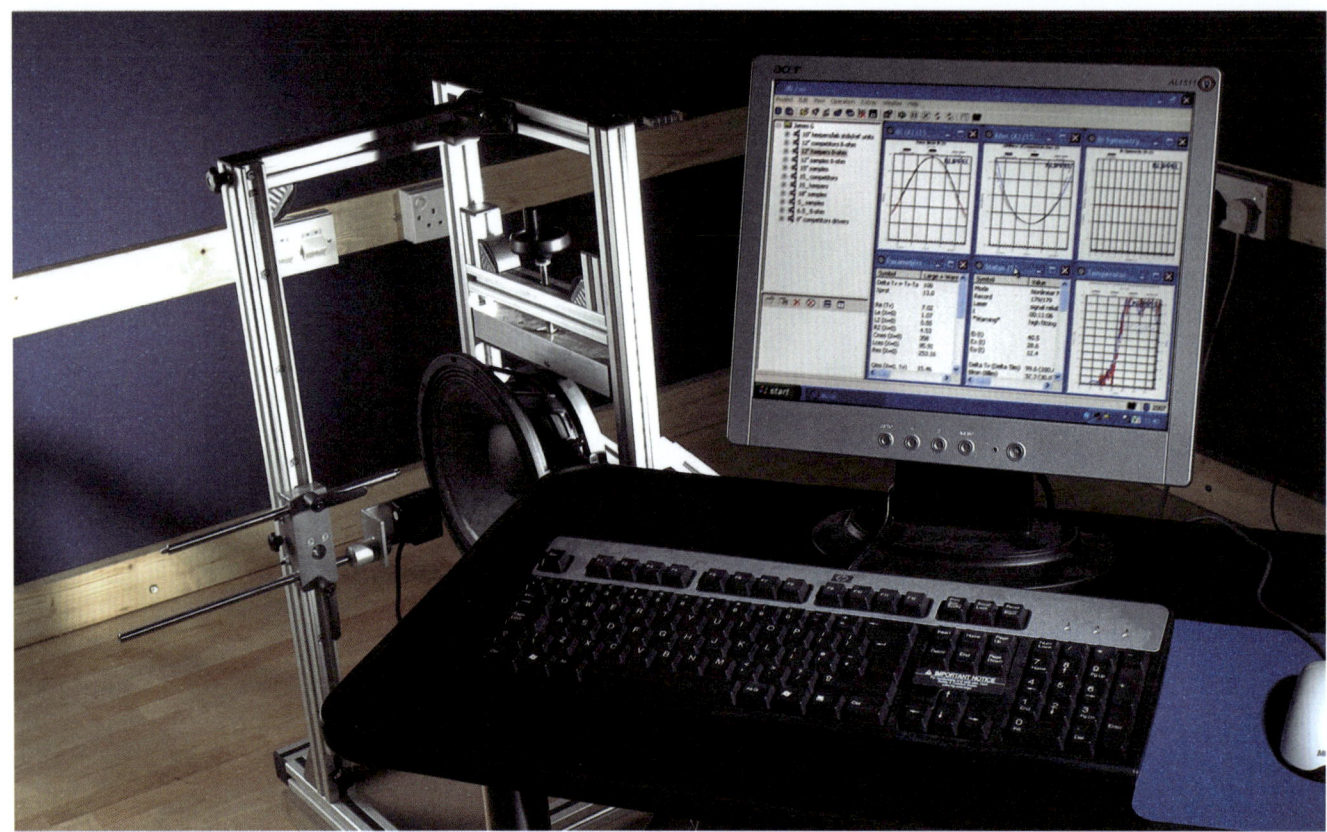

A sustained investment in development tools helped facilitate pro audio transducer development.

What followed was a realignment of Celestion pro audio, with a streamlining and rebranding of legacy products like the Truvox pressed steel woofers along with the addition of new ranges. An updated line of cast frame LF loudspeakers, intended to supplant the retail-led Frontline woofers, and as a nod to the quirky and clever inverted magnet NTi drivers, they were branded FTR: Ferrite Technology Rear Mounted. As the name implied, they had a conventional ferrite magnet mounted at the back of the driver, opposite to the NTi, exactly where you'd expect a magnet to be. No frills and no gimmicks, but they worked well. And they began to gain Celestion further traction as a credible pro audio transducer manufacturer.

More than anything else, though, it was compression drivers, built on the foundations of Mark Dodd's development work begun in the late 1990s, that formed the tip of the spear that enabled the company to begin targeting a serious reputation in professional audio sound reinforcement.

The transition away from finished systems encountered fewer bumps in the road in the United States than it had in the UK—where the range was embedded in the hearts and minds of many loyal customers; the shift there was altogether more seamless and paid immediate dividends. According to Andy Farrow, 'When we announced in 2006 that systems was done it opened the final door for us to QSC and a number of other high-profile PA partners, and the company's reputation in that market began to take off from there.'

The FTR speakers became a popular choice for OEMs.

By the time Celestion approached its 90th birthday, the long-term strategy—thinking in five-year cycles rather than bite-sized chunks, and emphasising the pro audio business—had proven to be a winning one. By then, the company was manufacturing an average of 1,000 compression drivers a day, a fact made possible in part thanks to the massive capacity of the GPE factory in Huizhou, China. Pro audio transducer sales to OEM companies at this point were generating 65% of the company's revenue, practically double that of the still extremely popular and successful guitar loudspeakers (guitar speaker sales had also grown, illustrating just how positive the response from the pro audio world had been). Celestion was on course to become arguably the biggest branded manufacturer of compression drivers—with Wood predicting that further down the line sales to the sound reinforcement sector would grow still further.

Much of this success had been possible only because of Gold Peak's unceasing support, as they worked with shareholders to make available the capital investment necessary to plough this new furrow of fertile business. The supply line of cash was invested and assiduously dispersed into the development of software tools, as well as engineering, sales and marketing muscle. Most importantly, Celestion had now won the trust and confidence of a global marketplace.

CELESTION

GUITAR LOUDSPEAKERS

THE VOICE OF ROCK & ROLL

18 The Voice of Rock & Roll

Once open, the Claydon factory immediately became a production hub for the Celestion Blue, reissued in the early 1990s and now considered the flagship guitar speaker. Also built here were custom special speakers, including several well-established models like the iconic G12 Vintage, the Wolverine and the G12B-150, which were all shipped directly to the Marshall factory in Bletchley, near Milton Keynes.

It was Marshall, again, whose continued desire to pay homage to the tones of the 1960s led to the development of the Heritage Series. This began with the creation of two 'new old' guitar speakers that would prove important over the next few years, as well as being Celestion's most considered attempt to recreate original 60s Greenback speakers in more than 20 years.

The brief was to recreate the G12M speaker for use in a newly revitalised Marshall 1974 and the G12H (the version with the 55 c/s low-resonance bass guitar cone) for use in a reproduction of the hand-wired 4x12 cabinet initially made famous by Hendrix. All built to a recipe that would enable the speaker to sound (and look) as close to the original 1960s versions as could be made in production quantity.

Paul Cork worked on the project as engineer, and recalled borrowing some old equipment from Marshall's museum of vintage amps. Examining those early Greenbacks, he noted many small differences compared with the more modern version of the G12M and G12H. To get close to the early designs meant researching adhesives to find a 'soundalike' formula and reverse-engineering the voice coil former. As with previous tonal re-creation projects, achieving an exact match of the cone was never going to be possible: instead, weeks of cone trials were undertaken in an attempt to identify the closest possible formula, to achieve a sound as near as possible to the benchmark tone.

As an additional tweak, the G12M was slightly under-magnetised compared with the contemporary G12M model, resulting in a more open tone that was a touch less aggressive, in-keeping with the well-worn sound of vintage speakers. Both G12M and G12H were given green cans and age-appropriate 'pre-Rola' labels, complete with Thames Ditton address. Production began in Claydon in mid-2005 and immediately began shipping to Marshall, with retail packs made available to the distribution network by the end of that year.

The Heritage Series G12H (55) guitar speaker.

For some, nothing will ever sound quite like a 'pre-Rola' Greenback, and there may be reasons for that: age and environmental factors have a sonic influence, and there were materials and manufacturing processes used in the 1960s that are no longer employed, all individually small changes that together have an incremental influence on speaker tone. For the guitar speaker-buying public at large, though, and for Marshall in particular, the Heritage Series was a great success.

The decision to build in the UK proved important too, as it became clear that Claydon was to have a role as a specialist factory with the ability to manufacture 'off-line' (hand-built without the need for a production line) product. This proved the perfect foil to the Chinese factory, which was entirely geared up for large-volume mass production.

It was also the first time that a low-resonance G12H had been offered since the mid-1970s, which for many was an exciting addition to the range. So popular were the sound and the concept of the Heritage Series that two further models were added over the next three years: another variant of the G12H (with 75 c/s lead guitar cone) and a reworking of 1978's G12-65, which had achieved near legendary status as a 4x12 speaker in some quarters with its late 70s rock vibe. Returning in 2007, it offered a guitar speaker tone unlike anything else in the range at that time.

For the guitar speaker-buying public at large ... and for Marshall in particular, the Heritage Series was a great success

Digging for Gold

As well loved as the Celestion Blue had always been, a common complaint was that the rated 15W power handling was too low. The internal policy had been to 'leave well alone', and not to interfere with the legacy of this iconic speaker. But in 2006 a decision was made to develop a higher-powered alnico speaker, to sound as close to that magic 'Blue' tone as possible, but with a little extra capacity.

Almost as difficult as the product development itself was deciding what colour to make it: in the end, Aztec Gold was the decision, and the first new alnico guitar speaker in over 40 years was launched. The UK-built Celestion Gold boasted a 50W rating, thanks in part to a cleverly reinforced voice coil former, and was officially launched at 2007's Winter NAMM Show.

How similar the Gold and Blue sound to each other is, as with all guitar tone discussions, a point of some debate. The Gold certainly had the texture and familiar dampened attack of an alnico speaker, with perhaps a more rolled-off treble in comparison to the chiming, bell-like Blue. Nevertheless, the sweet alnico tone was well-regarded enough to win a reviewer's award from *Guitar World* magazine and become country music star Brad Paisley's guitar speaker of choice: all in its first year of existence.

As the guitar speaker range continued to expand, it was heavily marketed once more, with a sustained advertising campaign in principal US magazines (the centre of gravity for sales). Although these were 'consumer' magazines, the ads were designed to create a pull for Celestion-loaded amps and cabs, reminding players of all kinds about the Celestion brand and its indelible link to the amps and players they loved. At the same time, a roster of endorsing artists was being cultivated and a broad spectrum of guitarists, from stadium headliners to studio session legends, playing everything from classic rock to metal, blues, country, funk, pop and jazz, were invited to tell us what it was they loved so much about Celestion and join on as a 'Partner in Tone'. More than anything else, this underscored that Celestion was THE go-to guitar speaker—just in case you were in any doubt.

Around the same time, the *Guitar Hero* computer game was launched, and as that franchise rose to massive popularity, it seemed that guitar music did too, as air-guitaring gamers were moving from virtual to actual guitars so they could rock out to their favourite tunes for real.

Right: The Celestion Gold, 50W alnico guitar speaker.

Left: As advertised by Brad Paisley.

The year 2008 was a high-water mark, with more guitar speakers sold than any year previously. The number of amp brands, both big and small, expanded, as new names joined steadfast and long-term partners. In the UK of course there was Vox, Marshall, Orange and Laney; notable European makers included Hughes & Kettner, Engl and Diezel; with US builders including Fender, Mesa/Boogie and, after a hiatus of more than 20 years, Peavey. There were many more besides, from exciting new brands like Blackstar, who had recently emerged in the English Midlands, to those offering the new digital technologies, including the rapidly expanding Line 6 from California. Boutique and artisan makers were building and tweaking cool and unusual amps in garages and small workshops; it truly was a boom time for guitar-playing gear.

Autograph Hunting

The introduction of the Heritage Series proved particularly timely when US salesman Rick Skillman fielded a chance call from Fender while driving down the I-405 freeway in California. The request was for samples of the recently launched G12M speaker, and the project was a collaboration between Fender and guitar legend Edward Van Halen, who together were to be introducing a new brand to the market, named EVH after the man himself.

Left: The G12 EVH signature guitar speaker.

Right: As advertised by Eddie Van Halen.

Lifetime Van Halen fan Skillman was in seventh heaven when, a few short months later, he and boss Andy Farrow were invited to visit Ed's large and secluded LA home to discuss the project and ended up spending much of the day at the mansion, also the location of the legendary 5150 Studios.

'Ed was passionate about gear and talked in detail about how his guitar rig worked, and how back in the day he had a very minimal setup, so had to improvise and make do with what he had which actually made him a better player,' remembered Skillman.

Expecting a lengthy process leading to the development of a new custom speaker, the pair were pleasantly surprised when Ed simply stated, 'No, I just love Greenbacks, so that's what I want, only with my artwork. I'm a 100% Greenback guy, and it's a big part of my sound, and we want to offer products that are exactly like what I use.'

The choice was between the newly released Heritage Series G12M, or the standard G12M version (reissued in the early 1990s and being built at the time at the GPE factory in China). In the end, the consensus was that the Heritage model would have been sonically closer to his original speakers from the first Van Halen album.

With the unique EVH-style stripes on the label, which, it turned out, were perfectly suited to a black rear can, the G12 EVH became the first signature product in the history of Celestion (not to mention the first 'Blackback' to be produced in 30 years). The speakers were loaded into the EVH-branded 4x12 cabs, and with the creation of subsequent custom speakers in later years it's a collaboration that remains highly successful, even after the great man's passing.

With the development of new guitar speakers, ideas came (and still come) in two principal ways. The first was when a customer would come with an idea and then work together with Celestion to achieve a shared aim. It had been a method used successfully with Marshall, Vox and Fender and many other iconic brands for decades.

The EVH speaker was devised in this mould, principally led by the customer, as were some later signature products. Another of these, developed together with shred legend George Lynch thanks to a mutual partnership with US amp manufacturer Randall, resulted in the development of the 'Lynchback', a hot-rodded G12M with an alternative voice coil that added extra punch to the rhythm sound. Another signature partnership, which came with a Fender logo, was Joe Bonamassa's JB-85 speaker that took a new cone to a Classic Lead 80, perfect for reproducing his smooth and deep blues tones.

The second route has always been based on a long and intimate knowledge of the guitar speaker market, born from decades of collective experience. By the 2010s it had become clear that 4x12 cabinets were considered too heavy and unwieldy, and the market now wanted smaller cabs. Whether it was because the average age of guitarists was increasing, or it simply reflected a more practical frame of mind, the call was out for somebody to build a speaker with the much sought-after vintage tonal characteristics but with a heftier power rating, to remain usable with the most muscular amplifiers. As in the 1970s and 80s, the evolution of the guitar speaker was again being driven by power and with a deliberate focus on sounding like the 'good old days'.

In essence, what the guitar-playing world seemed to be asking for was a 'high-powered Greenback'. In answer, the Celestion engineering team formulated a voice coil that delivered the heat resistance required for higher-powered speakers, without adding the tone-altering stiffness that some polymer coil formers were capable of doing. Something similar had been attempted before (notably with the G12T-75), but in Celestion's opinion this new take was tonally closer than ever, while simultaneously delivering a significant hike in power-handling capability.

In 2012, M magnet and H magnet versions were built, and after much debate and discussion were given cream-coloured cans—a nod to the short-lived period when cream rear cans were used on G12s built in the mid-1970s. All assembled in Claydon, they looked and sounded the part, and as soon as they were launched the new Creambacks were a hit.

The G12M-65 Creamback guitar speaker.

The tone of these speakers was outstanding. They were so popular that the range was quickly expanded to encompass a very high-powered alnico model (the 'Cream'), a 10in variant and, in 2016, a neodymium magnet model. The Neo Creamback came complete with a modified magnet assembly designed to attenuate the excessive treble that had plagued earlier neo guitar speaker models like the Century, due to the overabundance of magnetic flux generated by the extra-strong magnet.

That the Creambacks looked great and sounded great were two important reasons behind their popularity. But it couldn't be ignored that being UK-built was a strong advantage, so with that incentive, and after a little reorganisation of the Claydon factory, Celestion began to repatriate some of the premium guitar speakers that had been sent for production overseas some 15 years earlier. First to come back were the G12M Greenback, G12H Anniversary (the two original 1960s speakers that had been reissued in time for the company's 70th anniversary) and the Classic Lead 80.

That transition became a tipping point, with the Claydon factory now becoming the home of high-value, high-marque guitar speakers. Completely new models subsequently followed, including the high-power (150W) yet vintage-sounding Redback, the uber-high-powered (250W) Copperback and the smooth sounding, hemp-cone-equipped Hempback.

Left: The G12M-50 Hempback, hemp-coned guitar speaker.

Right: The 250W Copperback guitar speaker.

The Copperback had been 2018's attempt to create an ultra-clean neodymium speaker, a 'tonally improved' G12 Century, this time using the magnet assembly designed for the Neo Creamback which conveyed a smoother sound (the Century was almost universally considered 'too harsh'). It quickly found a place in one of the cabs from Quilter Labs, the guitar amp start-up established by QSC Audio Products founder Pat Quilter.

The latest UK product has been built in time for Celestion's 100th anniversary celebrations in 2024. Named the *100*, it has been voiced to be close to, and built to look like, the early silver alnico speakers of the late 1950s and early 60s. In a nod to modern guitar speaker design, however, the 100 has a more substantial power rating of 30W.

Celestion Digital

The launch of Celestion Impulse Responses (IRs) at the NAMM Show in 2017 raised a few eyebrows. At the same time, the company commenced the operation of the celestionplus.com online store as a means of bringing these officially Celestion-recorded digital downloads to the guitarists that were increasingly embracing new technologies, as well as marking the creation of 'Celestion Digital', which was set up as a sub-brand for this foray into the unfamiliar world of software.

Since the turn of the millennium, digital modellers had rapidly improved in technology, evolving from relatively simple combo amplifiers, with built-in EQ and effects, into advanced, standalone computing devices featuring sophisticated Digital Signal Processing (DSP) technology. By the

> One of the key enablers of this generation of digital music-making technology was the IR

early 2010s products such as Fractal's Axe-FX and Kemper's Profiler, alongside Two notes Audio's studio-based Torpedo load box and other technological trailblazers, had set the scene for a whole new category of digital product that sat comfortably alongside the DAWs (Digital Audio Workstation software) of the studio recording world.

One of the key enablers of this generation of digital music-making technology was the IR, a digitised 'snapshot' of a piece of equipment's acoustic behaviour. At this point IRs weren't particularly new or novel technology: they could and had been captured from anything that outputs or affects a sound signal, including, but not limited to, guitar speakers. However, with greater computer-processing power becoming cheaper and more available, the means to use IRs to emulate analogue music-making gear in *real time* was for the first time an affordable reality.

Generic IR captures of guitar speakers and cabinets were becoming readily available, and a growing cottage industry of small companies was emerging to produce these bits of software, including reproductions of Celestion-loaded cabinets. However, to Celestion's collective ears these IRs did not sound very representative of the speakers they were supposed to emulate.

The company decided it could do a better job digitally capturing its own speakers. That meant the best possible recording environment, the right expertise to help make the recordings and an optimised process to generate the software.

Decoy Studios, home to legendary mixing engineer Cenzo Townshend, was deemed an ideal recording space for the project (the fact that it was located in Woodbridge, just 20 minutes up the road from Celestion's base in Claydon, was an additional bonus). Research Engineer Andrew Harper developed a unique IR capture process, using industry-standard equipment. Entrusted with working the studio magic was the award-winning Mike Spink, who in addition to being an experienced sound mixer came with a reputation as a successful record producer, with credits including The 1975, Jake Bugg and The Gaslight Anthem.

Understanding studio sound was one thing, but Spink also knew how that sound would translate when played in a different space or put through a range of different equipment—which is exactly what would happen when audio professionals started using their IRs. He was also equally at home mixing live sound or studio recording and had been exposed to a vast range of different guitar sounds over the years.

Pete Thorn tries out Celestion IRs at Decoy studios in Suffolk.

As final proof of concept, Pete Thorn, renowned session guitarist and composer, was invited to visit the Claydon head office to take a listen and give an opinion. He was already well acquainted with the Celestion range of guitar speakers, but more importantly had been an early adopter of IR technology. Thorn instinctively understood what the company was trying to achieve and gave his seal of approval in resounding fashion.

However, with business to business being the heart of the company, and the majority of speaker sales going to manufacturers rather than through retailers, it seemed reasonable to apply that approach to the digital products as well. Celestion's head of marketing, Ken Weller, while overseeing the expansion of the celestionplus.com website set up to retail the IRs, also began to seek out manufacturers of digital music gear: from modelling amps to effects pedals, applying the tried and tested OEM business model with the goal of having them incorporate Celestion IRs in their products. Steadily Celestion Digital began to expand further, as new names like Positive Grid, Boss, TC Electronic and Headrush came into the IR fold.

The flexibility of digital amp technology and IRs was also revolutionising live performance, as many touring players traded in their bulky tube amps and 4x12s for a modelling amp that

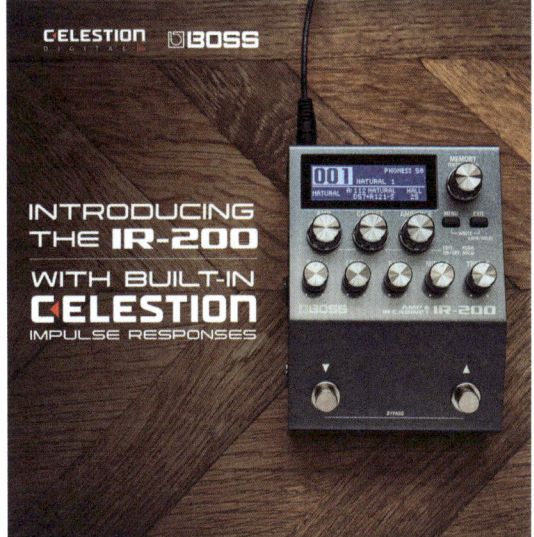
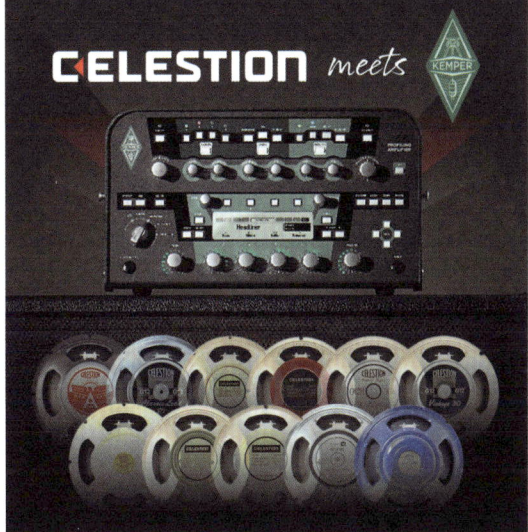

Left: The F12-X200 Full Range Live Response speaker.

Top and bottom right: Celestion IRs were quickly adopted for use with digital modelling hardware.

plugged directly into front of house to give a tone that was near enough (for most practical-minded gigging musicians) the same and plenty good enough for the majority of live venues.

This approach to live sound also spawned the Full Range Flat Response (FRFR) cabinet, essentially a dedicated monitor cabinet used to output the modelling signal—effective, but perhaps lacking in the musical character that a good backline would have. Spotting a further opportunity, Celestion introduced a Full Range *Live* Response speaker, specifically for use with digital modelling and profiling amps. The speaker was designed to get the best out of digital tone emulation, providing a broad-bandwidth, flat-frequency response, but adding just enough colouration to give the tone some life, which critics felt was missing from the first generation of FRFR products.

Digital was working out well for Celestion, and the next development saw the launch of Celestion-branded software, designed to work with the new and improved generation of IRs developed in-house: Dynamic Speaker Responses (DSRs).

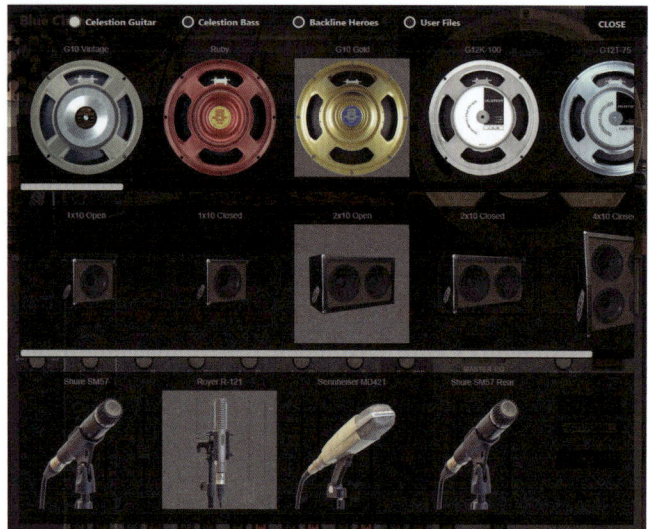

Screenshots of the SpeakerMix Pro DAW plug-in.

The original plan to create Celestion's own IR loader—a piece of software that enables an IR to be used with digital recording software (known as a DAW)—took on fresh momentum when, after working with a software partner on early designs, the project evolved into something much more complex and all-encompassing: a plug-in (a piece of add-on software for the DAW) called SpeakerMix Pro.

This software enabled the recording guitar player to virtually apply multiple speaker tones to a signal, while removing real-world speaker-recording problems like phase matching and incorporating a host of mixing desk functionality. Focusing specifically on speaker tone as it did, it was a unique concept.

While SpeakerMix Pro offered unique functionality, it was the companion DSRs that were the real ground-breaking innovation. 'While any IR can faithfully emulate the sound of a speaker, it does not respond well to the dynamics of playing,' explained Research Engineer Joe Elsom. 'Whether you hit the guitar strings hard or soft, the loudness may change but the underlying speaker tone remains the same. DSRs overcome that by incorporating the non-linear behaviour of a speaker in the real world to deliver much more nuance.'

'Every change in input level and style alters the "feel"—the harmonic response of the loudspeaker—which is what some say is lacking in the world of digital tone. By capturing these nuances, it's possible to add more "analogue realism" to the digital tonal model, incorporating dynamic elements that might otherwise be considered missing.'

The simple mechanical device given new life all those years ago by Les Ward might seem light-years away from the complex algorithms of digital modelling software; nevertheless, the tone remains reassuringly familiar, giving the same Celestion voice to guitar players across the decades.

Partners in Tone

The Partners in Tone guitar speaker endorsement campaign began in earnest at the beginning of the 2000s, and over the next few years it brought together a stellar line-up of famous and well-regarded players. It started slowly, with the recruitment of two very different players: Gem Archer (guitarist with Manchester band Oasis and friend of the Celestion sales team at the time) and Yngwie Malmsteen, the king of shred guitar and a Marshall player through and through. At the time, Yngwie told us, 'I like the Celestion G12T-75 speaker because it is very fluid and complements the violin-like tone and feel of my guitar playing. I have used Celestions since the early days of my career in Sweden.'

Arguably the seeds of the programme were sown as far back as the 1980s, when relationships were forged with the likes of Queen's Brian May in the UK. Stateside, Brian Coviello set about trying to make the company image a bit more rock 'n' roll as retail guitar speaker sales started taking off, actively seeking out guitarists and finding John Jorgensen (part of Elton John's band), Brad Whitford of Aerosmith and Billy Gibbons of ZZ Top, taking every opportunity to meet players at the annual musical instrument makers' and merchants' convention, NAMM.

However, it was when Slash officially joined the family in early 2004, and new marketing recruit John Paice was tasked with expanding the roster still further, that the programme began to earn its wings. As Slash himself told us, 'The first time I heard Celestion speakers I was hooked. They encompass everything that I'm trying to achieve sonically. I've never had to use anything else.'

By 2005, Paice was joined on artist relations duty by incoming salesman Rick Skillman, himself a gigging guitar player who set about bringing celebrated country music players, not often known for playing through Celestion, from the California and Nashville scenes and later from Austin and its environs. One long, enduring relationship was with guitarist/songwriter Brad Paisley, who, with 12 albums to date, can boast three GRAMMY® awards and 15 Country Music Association Awards. His favourite speaker is the Celestion Gold.

Other leading country music guitar slingers included Clayton Mitchell and Zeke Clark from Kenny Chesney's band, as well as Coy Bowles and Clay Cook from the Zac Brown Band.

Before long, the artist roster was populated with guitar heroes and bona fide legends from the 1960s through to the present day, all willing to pledge their fealty to the cause, for no other reason than the love of their speakers and how they sounded. Big players, it turned out, were fans of the brand and the Celestion sound, and were only too happy to join the extended family. Tony Iommi of Black Sabbath, for many THE original Heavy Metal guitarist, has been using used G12H speakers with his Laney cabinets since the first Black Sabbath album. He told us, 'Celestion is the voice of rock. Always has been. Always will be.'

All three guitarists from legendary Iron Maiden became endorsing artists evidenced

They're iconic and the name Celestion literally puts you in another zone

Joe Perry, Aerosmith

by repeated mentions in the band's coveted album sleeve notes. With Heavy Metal and a diverse range of related sub-genres perhaps at an all-time high during the mid-noughties, guitarists from a wide array of bands at that time were enthusiastic to sign on as a 'Partner in Tone'. Such luminaries as Adam D and Joel Stroetzel from Killswitch Engage, Jon Donais and Matt Bachand from Shadows Fall, Mike Spreitzer from DevilDriver, Bill Kelliher from Mastodon, and Paul Waggoner from Between the Buried and Me. Michael Amott from Arch Enemy told us, 'With my cabs loaded with Vintage 30s I am ready to deliver the full metal attack!'

Later in his career, Peter Frampton became a devotee of the Vintage 30, and that speaker has been part of Steve Vai's Carvin set-up since the late 1980s. According to Steve, 'Celestion speakers speak honestly and clearly. They enable me to hear what I'm thinking.'

Blues legend Robben Ford told us that he'd spent many years buying secondhand G12-65s because that was HIS tone, so he was extremely happy when Celestion reissued that speaker as part of the Heritage Series, and uses that speaker to this day.

Joe Perry of Aerosmith, a long-time Greenback player who's also known to experiment with Creambacks, kindly told us, 'they are beyond the mechanical fact that they are speakers. They're iconic and the name Celestion literally puts you in another zone.' And the whole project seemed to come full circle when Eric Clapton agreed officially to join the Partner in Tone family.

Sometimes endorsees came to Celestion through amp manufacturers and some players have reached out to us directly. Sometimes an endorsement contact would come from a more unusual place: a chance meeting with Angus Young's then guitar technician at a motorway service station, the late Geoff 'Bison' Banks, led to a close and lasting relationship with AC/DC. Of course, the Young brothers were Marshall players from day one, so Celestion was already part of their tonal DNA, with Angus playing through G12Ms and his much-missed brother Malcolm through various G12Hs, later switching to Vintage 30s. (Current rhythm player Stevie Young favours the Classic Lead 80.) As AC/DC prepare to return to the road for their 50th anniversary they will do so fully Celestion-loaded.

Opposite top left: Slash.
Opposite top right: Steve Vai.
Opposite bottom: Tony Iommi.

19 A Professional Audio Future

The guitar speaker business was clearly in the rudest of health, however, the glittering prize of Nigel Wood's strategy, set soon after his arrival, was to be further and faster expansion into the professional audio market.

Fundamental to this was the development of a more comprehensive range of transducers with an increasing focus on high performance; products intended to appeal to the high-end sound reinforcement customers dominating the sector. By the middle part of the 2010s, with more product available and a growing reputation for collaborative working, this mindset had begun to pay dividends.

The range expansion had begun with HF compression drivers initiated by Mark Dodd's CDX driver, which at the start of the new century had been re-engineered and renamed as the CDX1-1745. By 2003, Dodd's compression driver research had generated a paper presented to the Acoustical Engineering Society (AES), concerning the application of advanced finite element modelling methods to the design of these devices. It was a field in which Dodd, and the company at large, was gaining a considerable amount of expertise.

The work set out by that paper had already yielded the fundamental concepts for the development of three lightweight, neodymium magnet HF devices designed to be forward firing, with the phase plug coupled to the convex side of the diaphragm, which maximised the output Sound Pressure Level (SPL) of the device across its full output band, while keeping distortion low.

Turned from theory to finished product by Paul Cork, they featured a stiff and light aluminium dome (thought by some to be sonically superior to titanium). Early mainstays of the Celestion compression driver range, the CDX1-1415, 1425 and 1430, were well-received and were sold to a number of important customers in the battle to become better recognised as a pro audio component supplier of significance.

Opposite: The Celestion team outside the Claydon factory.

Being able to successfully build these highly sophisticated devices gave the company a definite competitive edge

From here, the CDX range was quickly expanded. This was due to the rapidly acquired expertise of the R&D team, who were becoming increasingly adept with the in-house software Celestion were developing for processing finite element calculations, and to the manufacturing expertise of GPE, which had the necessary skill to accomplish the precision fabrication of these complex and very small devices.

As Wood reflected sometime later, 'Compression drivers are complicated to make, so not everyone can do it. But we've not just worked out how, we know how to do it well. That complexity is our friend because the barrier to entry is high.' Being able to successfully build these highly sophisticated devices gave the company a definite competitive edge. It was an approach that was beginning to lead to success.

A further AES paper on compression drivers in conjunction with KEF's research engineer (later Vice President of Technology) Jack Oclee-Brown, and presented by Dodd in 2007, began a further line in CDX development which led to another unique design, dubbed the 'deep-drawn dome'. They had taken the compression driver's diaphragm, which had been roughly the same shape since the inception of the technology in the 1950s, and made it taller for increased stiffness. The new-style diaphragm was paired with a fundamentally redesigned phase plug—phase plugs are crucial for the effective operation of a compression driver—which was the real breakthrough of the design. The result was an output of much lower distortion all the way to the upper reaches of the device's frequency response.

It was an innovation that enabled the company to build a unique, ultra-low distortion, larger-format (3in voice coil, 1.4in exit) device to complement the smaller 1in exit drivers that were the current mainstay of the Celestion range, widening suitable application areas and hence potential customers. Audio purists loved the extreme clarity of the output, while traditionalists lamented the lack of HF 'ping' that they were used to. In the spirit of collaboration, Celestion offered more traditionally built options as well, and the CDX range continued to grow successfully.

Throughout the 2000s co-operation with sound reinforcement companies grew, and associations were cultivated with a broad range of organisations in Europe, the Americas and further afield. One such company was Renkus-Heinz, which collaborated with Celestion in the

The CDX1-1430 neodymium magnet compression driver.

mid-2000s to develop a custom LF unit, while Eastern Acoustic Works had begun using a 15in FTR model in one of their popular subwoofer cabinets, later turning to the forward-thinking and forward-firing aluminium-domed CDX1-1425 compression driver when specifying an HF component for their ambitious Anya line array.

Founder of QSC Audio Products Pat Quilter had initially come across Celestion as a young electronics engineer fixing his brother's guitar rig in the early 1970s, and when the company came calling in 2006 had thought Celestion pro audio speakers an interesting proposition. 'We actually came to Celestion pro audio through their parent company, Gold Peak Electronics,' explained Quilter. 'I thought they were building products with real engineering depth. They were also willing to work with us to develop custom drivers.' It was the beginning of a close partnership with the Californian company that continues to thrive.

By now, the company was actively identifying other areas where growth could be achieved, and one key strategy was to occupy market segments where only a limited selection of good-quality products was available. This could be because the application was yet to be fully recognised as an opportunity (as was the case with compact, full-range drivers for steerable column arrays), or as a result of the market segment requiring technically sophisticated devices that were difficult to build.

A large format compression driver (*top right*) with deep drawn diaphragm (*bottom right*) and associated 'maximum modal suppression' phase plug (*left*).

One such complex device was the coaxial loudspeaker, which combined LF and HF drivers and was used for applications that demanded closer time alignment between bass and treble for improved signal coherence, or simply where more compact cabinetry was needed. As was so often the case, the inspiration came from an OEM opportunity, in this case Martin Audio, and after a lengthy development phase a complete range of coaxial drivers was produced, destined for the company's forward-thinking CDD (Coaxial Differential Dispersion) range.

Martin Audio, a long-established and very prominent sound reinforcement manufacturer, had worked with Celestion as far back as the 1980s, when it had specified a customised version of the 1000W, B18-1000 driver for use in several subwoofer cabinets. In the 1990s, Martin Audio had launched several products containing Celestion drivers, such as an 8in bass/mid driver which was used with the popular EM26 installation cabinet.

A particularly significant collaboration between the two companies was the ICT300, which featured two specially designed 10in coaxial drivers, with the LF elements supplied by Celestion. These units also incorporated an HF diaphragm developed by Celestion alumnus Boaz Elieli, based on a 1980s HiFi design. The diaphragms featured no voice coil, but instead were energised by a process of inductive coupling using the magnetic field generated by the LF coil.

Around the mid to late 2000s, Martin Audio and Celestion picked up their relationship once more. This was in no small measure because of Celestion's renewed determination to expand its presence in the pro driver market, its expertise in, and experience of, FEA and, just as significantly, its willingness to collaborate. Martin Audio's R&D director at the time, Jason Baird, specified a Celestion 6.5in driver in the popular DD6, with the two companies later cooperating on the design of the highly bespoke LF and HF drivers for Martin Audio's MLA Mini ultra-compact line array.

Current Martin Audio Managing Director Dom Harter had been a regular user of Celestion transducers when overseeing R&D during a tenure at Turbosound (from 2002 to 2012): 'The [audio industry in the] UK has a certain midrange philosophy and what we were really looking for were companies who had the skillset to make some difficult midrange drivers.' Harter became aware that woofer specialist Neville Ryan had returned to Celestion as OEM Engineering Manager after leaving another transducer company, Precision Devices, which was part of the same EdgeTech Group as Turbosound, and therefore its principal supplier.

'Because nobody really made these specialist midrange drivers we were attracted in that direction. There was now someone who had done it for Turbo for years, we could talk the same language ... and that led to Turbo doing a lot of business with Celestion.' In fact, Celestion developed some very unique midrange speakers for Turbosound during that time, including the 10in driver for the TA-890 Aspect system.

On taking over the helm at Martin Audio in March 2016, Harter, being familiar with the company's capabilities, quickly made use of Celestion's in-house strengths, including the FEA resource and expertise: 'The experience they have using those tools is pretty advanced. Therefore, if there is a process of research embedded in the design phase, we'll generally go to Celestion, partly because they have good modelling equipment, and also because of the relationship we've developed.'

'Many suppliers will look at the Thiele and Small parameters, grab the nearest thing off the shelf that kind of matches what you are looking to achieve and then play with it until you get you what you want. But Celestion operate a little bit differently, and are great when we want something a bit out of the norm and unusual. The fact that they are based in the UK is good news; we have a balance between volume capacity in the Far East, with engineering experience, stock and service in the UK so it works well for us.'

'We've honed down our suppliers to people who have engineering skills, decent quality and operate to a standard appropriate to our brand.' The reality today, he concludes, is that the re-engagement with Celestion over a decade and a half ago has yielded rich fruit, notably with the current and highly successful CDD range: 'The best testament to Celestion is that their products are now in half our optimised line arrays—and as we replace the DD6, 13 years after its launch, we will again do so with Celestion.'

Left: Computer Aided Design (CAD) render of new style coaxial speaker, launched late 2023.

Right: Exploded view of an FTX0820.

The advances made as a result of the coaxial project made possible the standard FTX range, which featured a 'common motor', where one magnet assembly supplied the magnetic circuit for both the LF and HF components of the driver. While not a new concept, the technology was adapted and optimised by Celestion's engineers, again assisted by the sophisticated in-house design software, which in this case enabled the critical balancing of magnetic flux for HF and LF voice coils.

What emerged was a clever and elegant solution to remove significant size and weight from the final products while extracting the maximum performance possible, ultimately securing Celestion a market-leading position in yet another professional audio product segment.

Finite Elements

When finite element methods was adopted in the late 1980s as a means of looking at loudspeaker behaviour, it was still an esoteric and academic discipline, arguably better suited to scientists dedicated to analysis rather than to engineers who were more interested in hitting a performance brief than solving differential equations for fun. Additionally, it demanded a lot of expensive computer-processing power as well as the patience to manipulate software still in the early stages of development. However, Celestion had recognised its importance and usefulness early on, and over the ensuing years had worked to understand the processes involved. The company overcame difficulties in the implementation of finite element calculations within the available software and adapted the available technology, making it work successfully for developing loudspeakers.

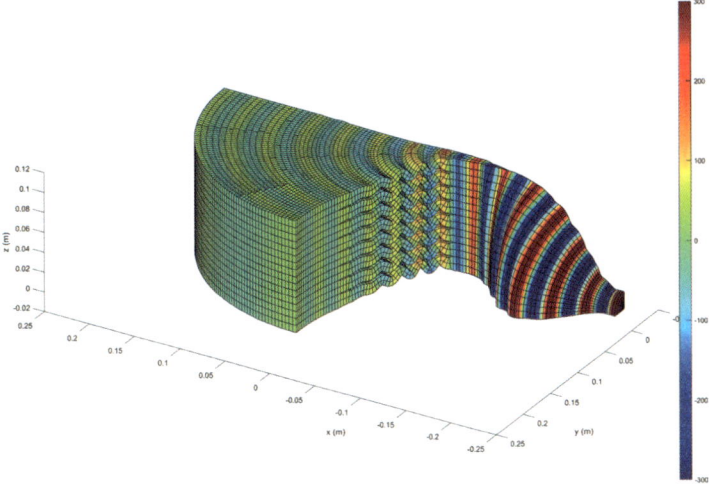

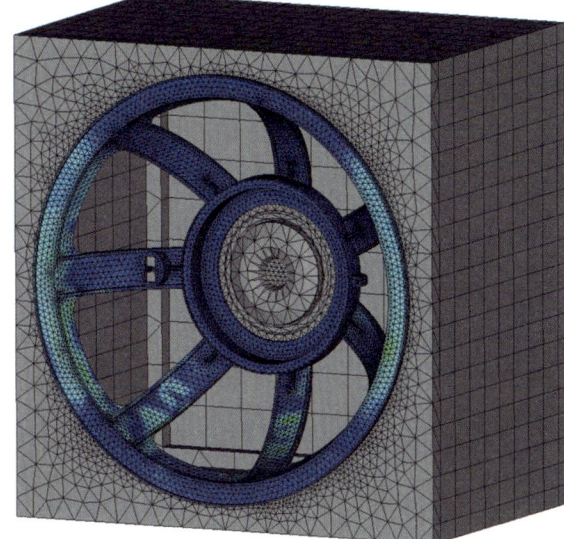

Left: An FEA simulation of a line array HF waveguide.

Right: Finite element methods can be used to optimise the strength of a chassis housing and its outer packaging.

By the 2010s Celestion had created a fully bespoke, highly sophisticated and, above all, very engineer-friendly piece of loudspeaker design software. It was a system that could mathematically model the whole loudspeaker system mechanically, electrically, thermally and acoustically, accurately predicting its behaviour purely from calculations and delivering an insight into how each part of the interconnected system influences the other and hence the product as a whole.

In more recent years, the analysis system was extended to look beyond the loudspeaker to include horns: another crucial component for the operation of HF compression drivers. By incorporating an additional skillset and applying it to the FEA system, Celestion's engineers were able to offer another customisable product that would ordinarily be difficult to design well.

The expertise gained by developing this system over several years while working on complex and varied projects, coupled with the power of the software that has been created as a result, brought about a situation where a new speaker could theoretically be designed from scratch, without the need to build a test prototype. Just as critical to the process was an in-depth knowledge of all of the potential materials that might be used in the design process—a knowledge that had been painstakingly compiled in a detailed database of measurements that are vital to accurate mathematical modelling. The designer could now piece together each component virtually and be confident that the finished outcome would match the speaker's performance in the real world. That made it a formidable tool and a genuine point of difference for the company.

Left: The Axi2050 wideband compression driver.

Right: CAD software representation of the main Axi2050 components.

Made in Britain

It was at the 2016 Prolight+Sound Show in Frankfurt that Celestion first unveiled its 'AxiPeriodic' technology in the shape of the prototype Axi2050 wideband compression driver. The Axi was specifically designed to outperform the mid/high compression drivers that were available at the time. These drivers, although reasonably effective, functioned in the same way as a coaxial driver, with two separate motors and individual diaphragms. This made them cumbersome and also demanded the use of a crossover circuit, with all the associated problems of combining two distinct signals into one coherent output.

Functioning as a single compression driver, the Axi2050 completely removed the need for a crossover, instead performing the role of two speakers in one, delivering a frequency range of 300Hz to 20,000Hz without the need for a crossover: very much a 'welcome to the future' moment for pro audio driver development.

Often bulky and expensive, crossovers can introduce phasing and other distortion effects when combining signals. This can negatively impact the coherence of the output, degrading sound quality, often just in the frequency range where the voices are. Taking away the need for a crossover takes away the performance problem, bringing a noticeable clarity to the Axi2050's reproduction. That the device was also designed to achieve this at the high output levels demanded for sound reinforcement was a real masterstroke.

The performance relied upon a 'heavily sculpted, circumferentially AxiPeriodic annular titanium diaphragm' (the name AxiPeriodic referred to the geometry of the diaphragm), together with a sophisticated phase plug, which was intrinsic to removing unwanted modes of vibration from the response, further adding to signal clarity.

Because of the sophistication of the design, it took another two years after launch for the AxiPeriodic to be brought to production, in a newly created and specialised manufacturing cell in the Claydon factory. 'AxiPeriodic got us into new OEM accounts,' commented Andy Farrow, 'including some very high calibre recording studios.' A sure sign that an innovative, risk-taking approach was yielding dividends in the form of new customers and applications.

It was a product that also opened the door to another exciting opportunity: the manufacture of high-end pro audio speakers in the UK. Since the near-wholesale transition of manufacturing to the GPE factory in Guangdong Province, Claydon had seen only the manufacture of guitar speakers, plus prototypes and small runs of specialist mid-range and very large subwoofers for one or two European customers.

One of those customers was Funktion One, co-founded by sound reinforcement systems icon Tony Andrews, who had begun working with Celestion during the mid-2000s: 'I'd been aware of Celestion for … like forever, and knew they were one of the first companies to get a Laser Interferometry device. But it wasn't until we [Funktion One] developed the smaller Compact Series in the mid-2000s that we started working with them initially taking advantage of their value engineering. We used them in the entire range—the 5s, 8s, 10s and 12s.'

'Nearly all of our drivers are special, and straight away [Celestion] allowed us to make a few modifications to standard product, for instance to the coil geometry. So it's gone really well, with good reliability and repeatability.'

Andrews worked initially with Paul Cork and later with Neville Ryan—both hugely experienced engineers. 'In fact Celestion probably have one of the best groups of loudspeaker engineers in this country. They are just so amenable to suggestion and discussion—their level of engineering understanding and reliability tests are just excellent.'

After the partnership had been so successful on the smaller commodity items, Funktion One turned to their current flagship [Vero] VX system, which was loaded from the ground up with Celestion: 'The four or five years prior to its launch we had worked together on special drivers for the VX including a custom 4in coil, 12in speaker. But where it really shone was that they brought some new ideas to our sacred midrange drivers. The suggestion was made and we tried it, and we evolved a better way of doing our midrange with them. So it's been a very successful collaboration.'

With the capability of UK pro audio manufacture now a proven concept, the seed of an idea had been sown. With changes to shift patterns and material sourcing, more efficient production processes and some good old fashioned furniture removal, it became possible to commission another full production line to be used for building customised variants of LF speakers. This

The mantra for this new range had become 'longevity of high performance'

move quickly unlocked opportunities with another high-marque tour sound company who had been in the sales crosshairs for some time and, just like that, the compact Claydon plant also became a hub of specialist pro audio manufacturing.

The coming of Covid-19, and the ensuing lockdowns in 2020 and 2021 that brought much of the world to a standstill, took the wind out of the pro audio industry's sails, as the annual carnival of music festivals and trans-continental tours ground to a halt. This was a glitch in the progress of Celestion's pro audio strategy, as customers had to call time on new product introductions, with many going into complete hibernation. Thankfully for Celestion, pro audio sales continued to tick over, but even more fortunately for the company, with so many people stuck at home, home music making skyrocketed and guitar speaker sales similarly escalated, ensuring the commercial stability of the company during what was for many a very worrisome time.

The hiatus in much of the co-operative OEM design work that had become the daily routine for the company also gave it time to carefully plan what should come next. Celestion was undeniably now a leading brand for pro audio compression driver and coaxial speakers, and it was time to build the same reputation for LF speakers once and for all.

With the occasional exception, such as the BX Series in the 1990s, or the NTi woofers at the turn of the millennium—both of which had been quirky and clever but had underperformed in sales terms—by and large, Celestion's woofers had the historical reputation of being capable if utilitarian. Good, solid performers that sold well and appeared in a wide range of brands around the world. It was recognised that a shift in approach was needed if Celestion wished to be viewed in the same light as the true market-leading woofers.

The company clearly had the R&D skills to achieve this, and a lengthy pedigree of recent pro audio designs with manufacturing partners had honed those skills for the development of LF devices, particularly in the period following the development of the CF Series begun in 2014 through to the creation of 2023's champion of performance longevity, the PowerProX18.

At the same time, Celestion engineers had one eye on continual and ongoing product improvement. One perfect example of this was the magnet assembly cooling system used for heavy duty subwoofer drivers intended to take a large amount of input power. A new approach to heat reduction used clustered groups of circular vents, each of a different diameter, routing controlled amounts of air around specific parts of the magnet assembly to maximise cooling.

Left: The precision assembly of suspensions for prestige LF drivers.

Right: A new robotically assisted production line in Claydon.

Senior Development Engineer James Gibbons had tested the concept mathematically using the design software and then visualised the airflow in the real world using a specially built 'smoke chamber'. This highly unusual method utilised smoke to trace the airflow through the magnet assembly to make sure that fluid flow around the metalwork was exactly where it needed to be to have the greatest effect. On completion it was considered by the Celestion team as the gold standard for speaker motor cooling, with the concept becoming known as Precision Tuned Venting. Once ready for production, it was implemented on prestige LF drivers that were beginning to be built in the Claydon factory.

Air cooling optimisation is far from unique in the loudspeaker world, but, by developing this novel approach, Celestion had made a significant improvement that had a real-world impact on the performance of its transducers. And it was the idea of constant incremental improvement that was to become the theme for the next phase of LF speaker production.

During the Covid-19 slowdown, a specialist programme was fast-tracked with the aim of looking at other marginal but significant improvements that could be made to woofers and subwoofers. Additional improvements to heat resistance and temperature management were augmented by changes to manufacturing processes, and the use of new materials that, when brought together, offered significant performance breakthroughs.

The mantra for this new range had become 'longevity of high performance', the idea being to produce a speaker whose performance was as close as possible to specification, even after hundreds of hours of use. After all, wear and tear is inevitable, but if it were possible to build in a number of features that could hold back the ageing process so to speak, the performance improvement could be impressive. It was a bold aim, but early test results were so impressive that, by 2022, the company began planning to invest in a state-of-the-art, robotically assisted production line to build the new family of products in the Claydon factory.

As a consequence, the Claydon shop floor area was significantly expanded and the commissioning of a brand new, full production line was scheduled for completion in late 2023. The strategy to build high-performance speakers in the UK on a large scale once more was in fully effect, with the plan and intention to expand still further as demand increases. With one eye on an official launch at the same time as 2024's Centenary celebrations, this range has been officially dubbed TSQ or 'Ten Squared' – a name intended to raise a smile among the more mathematically inclined: Ten Squared, or $10^2 = 100$.

That the financial capital was made available for this sizeable investment into technologically sophisticated manufacturing equipment was testament to Gold Peak's ongoing promise to support Celestion in its growth. In fact, the additional UK production capacity was essentially the third part of a three-prong investment approach in a little over three years, by the parent company, intended to secure long-term growth for the Group as a whole.

Earlier in 2023, GPE had completed a long-planned moved to a massive new design and manufacturing premises on the rapidly expanding outskirts of Huizhou, following on from the acquisition of a factory in Thailand intended for both electronics and loudspeaker manufacturing. This support is a solid foundation for Celestion as it continues to build for the future as an innovative British manufacturer with a global perspective. As that future beckons, the company intends more of the same: genuine innovations conceived by a capable and highly motivated research team, such as the recent design of a highly innovative (and patented) new approach used to create a wide directivity horn and wave-shaper that offers unprecedented sound wave coherence for the HF elements of line arrays. In 2024 this will be made real through Celestion engineering expertise and the advanced development tools that are now well established.

It's an approach that's in no small measure down to the organisation built by the current and longest-serving managing director, Nigel Wood, whose ability to see the most direct route through the day-to-day noise and clutter has ably steered the company for (at the time of writing) almost 20 years, by ensuring it sticks to doing what it has always done best. Or, as Nigel himself succinctly puts it, '100 years ago we started making speakers and 100 years later we are still making speakers.'

Celestion has 100 years of *bona fide* industrial heritage and a considerable legacy that runs unbroken through the annals of audio reproduction. Yet, rather than resting on those comfortable laurels, the company's focus is the ongoing journey to the next significant milestone. That's the type of long-termism that's at the core of the company ethos, firmly and steadily building for the future based on lessons learned from a significant history.

Ten Squared low frequency
loudspeakers, built for longevity of
high performance.

A PROFESSIONAL AUDIO FUTURE

Appendix I: The T-Book

The T-book is Celestion's catalogue of finished product part numbers. It's a vast list of every product sold (and a few that never were), compiled since the early 1950s. Below is a sample of its contents: it's by no means a complete list, there wasn't enough space to include all 6000+ entries! However it does contain products mentioned throughout this book as well as other products of interest and note. Over the years it has been active, not everything was dated and some of the descriptions are opaque or somewhat vague, but that's just part of the magic.

Key: PE – paper edge; CE – cloth edge; V/C – voice coil; D/D – dust dome.

T Number	Model	Date Issued	Ω	Description
0530	G12		8	As CT3757 sprayed blue with H1668 cover (Vox Blue).
0553	HF1300			HF1300 MkII ceramic magnet.
0555, 56				Colaudio II. Cabinet model mahogany, walnut.
650	G12		8	As T530 less cover, sprayed silver (Marshall Silver).
652	G12		15	As T650 but 15 Ohm (Marshall Silver).
655	Column LS			Fitted with 3x SD24 (T0007), 4x special 10in LS.
0703	G12		8	10 watt Jennings L/S as T530 but with G44 magnet assy.
1009	G44		15	G44 type. T0530 fitted cone assy. Magnet as P44.
1021	G18C		5	3in Coil, Jennings, twin coil, aluminium wire, ceramic magnet.
1087-90	G12		8	As T530, sprayed black, poly grey, dark grey, dark grey hammer.
1102	G15C	31/08/66	15	As T1074 golden sand housing, olive green cover.
1104	LS			Trawler talk back with transformer.
1134	G12		10-12	H1777 cone. Edge treat to first corrugated. CX2012 mag assy.
1172,73	Ditton 10			Walnut, teak HiFi cabinet assy.
1174	G12C		15	Fitted H1777-Cone guitar finish. Magnet ceramic.
1175	G12C		15	Light magnet. Housing unsprayed H1777-cone, guitar finish.
1178	G18C		15	Spray oyster hammer, fit Muller cone, 30 cps.
1220	G12M		8	Ceramic magnet group, plastic cover, golden sand.
1220-67	G12M		8	Heritage series G12M, 20 watt, 96dB, 1.75in coil, 75Hz.
1221	G12M		15	As T1220 but 15 Ohm.
1221-67	G12M		15	Heritage series G12M, as T1220-67 but 15 ohm.
1234, 1281	G12H		8, 15	G12H standard finish with cover. Heritage Series G12H(55Hz).
1268, 69	G12M		8	Jennings. Poly grey housing and special plastic cover.
1277	G12M		15	Standard edge finish 65-70cs. Marshall name plate.
1278	G12H		15	Standard edge finish 65-70cs. Marshall name plate.

T Number	Model	Date Issued	Ω	Description
1279	G12H		8	Jennings colour and Vox name plate.
1280	G18C		3	Free edge. Jennings colour and name plate.
1360	MH1000		15	Horn/Tweeter unit. 25w, 15 Ohm.
1361, 62	Ditton 15			HiFi cab assy. Teak, walnut.
1363	G12H		15	Standard G12M cone assy and colour. Heritage G12H (75hz).
1364	G12H		8	As T1363 but 8 Ohm. Also Heritage G12H (75hz).
1384, 85	G12L		15, 8	Standard colours, golden sand and green cover.
1420	MF1000		8	Mid frequency horn LS.
1454	G12S		8	Cone assy as T1088, Jennings colour chassis, standard cover.
1460	G12H		8	SP444 cone, standard edge treatment.
1482	G12H		15	50 watt, nomex coil former, standard colours.
1517	G12S		15	H1777 cone, edge treat, G12M cover.
1520	G12H		10	SP0444 cone, edge treat, tweeter SP0773.
1594	G12H		15	H1777-Cone, edge treat- twin cone - to handle 50 Watts.
1603	HF2000	19/6/68	4&8	SA3880-voice coil.
1604, 5	Ditton 25			HiFi cab assy. Teak, walnut.
1651	G12S		4	Bang & Olufsen steel chassis G12S magnet - experimental.
1757	G15M		16	50 watt, Alu dust cap, 003 cone. 55Hz.
1794	Ditton 25 ABR			Dynacord cabinet 18in LS - free edge ABR without damping.
1885	G12M		8	SA3422 coil, pulsonic 102/030 cone. Marshall.
1886	G12H		15	SP789 coil, pulsonic 102/030 cone. Marshall.
1900	Ditton 120	15/04/71		Ditton 120 Type Cabinet Assy (Bell & Howell).
1907	HF1000 Drive Unit	05/05/71	15	Complete drive unit - as (T1676) less flare (Dynacord).
1934, 35	Ditton 44	25/8/71	4&8	HiFi cab assy. Teak. walnut
2008	Powercel 15	16/12/71	8	Colour saffron & black - (res: 45c/s) - 125Watt.
2011-12	Ditton 66	03/06/72	4	HiFi cab assy. Teak, paldao.
2014-2015	County		4&8	Budget cabinet system teak, paldao.
2031	TELEFI			Teak - 5.5MHz.
2144-45	Ditton 44	19/2/73		Teak, walnut. Velcro changed to hedlock, redesigned front grille.
2158	HD500	15/5/73		Rubber rings glued with bostik 1755 - Bang & Olufsen.
2230	HF1300	19/4/74	15	Nomex - BBC spec. - BP closed - replaces (T1306).
2241	G12/50	25/6/74	8	Cone piston cambric edge, yellow housing black cover.
2251, 52	Powercel 15, 12		10	Chassis fire engine red - black cover.
2269,70	UL6	11/05/74	4&8	HiFi cab assy. Teak, walnut. 6PL + MD1000 + ABR.
2272,73	UL8	11/05/74	4&8	HiFi cab assy. Teak, walnut. 8PL + MD1000 + ABR.

T Number	Model	Date Issued	Ω	Description
2275,76	UL10	11/05/74	4&8	HiFi cab assy. Teak, walnut.10PL + MD700 + HF2000.
2277	G12H	11/05/74	15	Marshall. Housing sprayed red with black cover.
2322-23	D33		4&8	HiFi cab assy. Teak, walnut.
2372	Twin Horn System	08/12/75	30	Marshall - red - twin 15ohm horns on cast escutcheon.
2433	Hadleigh	03/08/76		HiFi cab assy. Teak.
2607-08	Ditton 22	27/10/76	4&8	HiFi cab assy. Teak, walnut.
2723-24	Ditton 15XR	13/5/77	8	Teak, walnut - Ditton 15XR.
2758	Constable Dedham	17/8/77		Modified Ditton 66. Grille hand woven full width of front.
2834,35	G12-65	03/07/78	8, 15	New power rating of standard G12M. PE. Silver D/D.
2836,37	G12-80		8, 15	New power rating of standard G12H. PE. Silver D/D.
2838-41	G12-65	03/09/78	8	Various permutations of D/D.
2954-2957	Ditton 551	25/04/78		HiFi cab assy. American walnut, elm.
3034-37	Ditton 662	19/06/78		HiFi cab assy. American walnut, elm.
3053-54	G12-65	22/6/78	8, 15	H1777-cone cambric D/D black, also Heritage Series.
3126-27	Ditton 442	15/11/78		HiFi cab assy. Black ash.
3211	Ditton 121	16/2/79		HiFi cab assy. Penang G vinyl.
3253-58	Ditton 332	13/3/79		HiFi cab assy. American walnut, elm, black ash.
3298	G12-65	30/8/79	8	For Marshall. As (T3053) but Marshall label.
3301	P1 Kit	09/03/79		Sound reinforcement cabinet system.
3357-59	Ditton 120	20/02/80		HiFi cab assy. Japan teak, Penang G, Black.
3369-70	Ditton 130	18/04/80		HiFi cab assy. Walnut, teak.
3371-72	Ditton 200	05/12/80		HiFi cab assy. Walnut, teak.
3382-83	Ditton 150	22/05/80		HiFi cab assy. Walnut, teak.
3393-95	Truvox 180, 230, 360		8	HiFi cab assy. Walnut.
3425	HF50	10/06/80	16	Horn treble unit.
3445	RTT50	12/10/80	8	Bullet horn.
3504-05	SL6	15/07/81	8	HiFi cab assy. Walnut, black.
3509	Ditton 3, 5, 8	13/8/81	8	HiFi cab assy. Walnut.
3513-14	SL6	29/09/81	8	HiFi cab assy. Rio-rosewood, Mexican.
3585	G12K-85		8	PE. Was G12-80. Now G12K-100.
3660	Ditton 240	12/10/82		Walnut - with CP metallic brown baffle.
3760	G12T-75		16	1.75in coil - Marshall - replaces T3054.
3771-72	S12-150		8, 16	Sidewinder 12in PE.
3779	SL600		8	HiFi cab assy. Dark brown.
3781, 82	G12T-75		4, 8	1¾in coil - Marshall - replaces (T3053).
3830-32	DL Series			HiFi cab assy. walnut.
3839-40	S15-250		8, 16	Sidewinder 15in.
3865-66	S12-150		8, 16	Sidewinder - cambric edge.
3903-04	G12		8, 16	Celestion Vintage 30.
3918-19	SL6S		8	HiFi cab assy. Rosewood, black.
3920-21	Concerto	25/09/85	8	HiFi cab assy. Walnut, black.

T Number	Model	Date Issued	Ω	Description
3933	System 6000	05/07/86		Subwoofer.
3937-38	Ditton 1		8	HiFi cab assy. Walnut, black.
3941-42	Ditton 2		8	HiFi cab assy. Walnut, black.
3943-44	Ditton 3		8	HiFi cab assy. Walnut, black.
3966	Ditton 88	11/12/86	8	HiFi cab assy. Walnut.
3969, 78	G12-80		8, 16	Special - 'Classic Lead' version.
3971, 79	G12-70		8	Special - 'Modern Lead' version.
3973	SR1	29/12/87	8	Dual speaker PA system moulded cabinet.
3974	SR2	01/06/87	8	B18-1000 plus cabinet.
3982	SRC-1 Controller			For SR1.
3983	SR3		8	
3989	G12-80		8	Special for Mesa.
4042	18in Martin Audio	09/08/87	8	Special for Martin Audio. Sprayed surround both sides. New V/C.
4070-71	SL6Si	28/4/88	8	HiFi cab assy. Walnut, black.
4072-73	SL12Si	28/4/88	8	HiFi cab assy. Walnut, black.
4074	SL600Si	28/4/88	8	HiFi cab assy. Dark grey.
4095	Concertino	28/6/88	8	HiFi cab assy. Black - similar to SL6Si.
4103	Yam 6.5	19/8/88	8	6.5in Mid Range Unit for Yamaha - power product.
4108-09.	Celestion 3		8	HiFi cab assy. Walnut, black.
4142	G12B-150	21/3/89	8	Special for Marshall - PE.
4145	Celestion 3000 (pair)	19/5/89	8	Ribbon tweeter. Black Ash Vinyl.
4148	Celestion 5000 (pair)	19/5/89	8	Ribbon tweeter. American walnut real wood veneer.
4151	Celestion 7000 (pair)	19/5/89	8	Ribbon tweeter. American walnut real wood veneer.
4204, 05	Celestion 700/SE	02/06/90	8	HiFi cabinet assy.
4208-09	Celestion 5	03/08/90	8	HiFi cab assy. Oak, black.
4248-51	BX Series, 15in	01/11/91	8	15in Prof Range (push terminals rubber tyre etc).
4267-68	Celestion 7	31/01/91		HiFi cab assy. Oak, black.
4269-70	Celestion 9	31/01/91		HiFi cab assy. Oak, black.
4271-72	Celestion 11	31/01/91		HiFi cab assy. Oak, black.
4276-79	Celestion 100	31/01/91		HiFi cab assy. Walnut, mahogany, black oak.
4301	SRi One	21/06/91	8	PA cab assy. Black.
4302	SRi Two	21/06/91	8	PA cab assy. Black.
4316, 18	AD15H	27/09/91	8, 4	15in Automotive Sub - BX15 housing - logo on D/D
4333, 69	Vintage 10	12/04/91	8, 16	1.75in V/C.
4335, 4416	Vintage 30		8, 16	Mesa.
4342	G12T Special	15/1/92	4	Marshall part No. S113.
4343-45	Celestion 300	28/1/92	8	HiFi cab assy. Walnut, mahogany, black. Transmission line.
4347	C10T-80	30/1/92	32	Special for Trace Elliot - kevlar cone - logo on D/D.
4355	G12 Gold	13/2/92	8	Special for Marshall - Marshall part No. S303.
4359-60	Celestion 15	27/2/92		HiFi cab assy. Oak, black.

T Number	Model	Date Issued	Ω	Description
4371-72	Celestion 3 Mk. 2			HiFi cab assy. Oak, black.
4373-74	Celestion 5 Mk. 2			HiFi cab assy. Oak, black.
4375-76	Celestion 7 Mk. 2			HiFi cab assy. Oak, black.
4388-89	Ditton 11	20/7/92	8	HiFi cab assy.
4390-91	Ditton 22	20/7/92	8	HiFi cab assy. Walnut.
4392-93	Ditton 33	20/7/92	8	HiFi cab assy. Walnut.
4394-98	Ditton Legend	20/7/92	8	HiFi cab assy. Black.
4427,36	Celestion Blue		8, 15	Std Celestion. Vox Blue (T0530) different R/Label.
4455	Vintage 30	15/2/93	5.3	Special for Trace Elliot as (T3903) but 5.3ohm coil. New R/Label
4500	V12	13/8/93	8	Special for Crate - green housing
4510	Celestion KR1	27/09/93	4	Dark Grey Finish (T4584 Black Finish)
4514-21	CR Series	20/10/93	8	Black carpeted plywood cabinet sound reinforcement system.
4533-34	G12H (30W)	10/11/93	8, 16	Standard Golden Sand housing. 70th Anniversary.
4535-40	CRi - 102M	15/12/93	8	CRi series, black textured finish, sound reinforcement system.
4586-87	Kingston (Inc Stand)	24/05/94		HiFi cab assy. Light or dark finish.
4612,61	G12 Heritage	14/10/94	16, 8	Special For Marshall.
4659	K10T-150 I.C.T	06/03/95	16	Special For Martin Audio. No rear label.
4674	KR8	22/05/95	8	Moulded plastic cabinet (front & back).
4679	CSW MK2	20/06/95		Powered Subwoofer.
4680	KR10	21/06/95	8	Moulded plastic cabinet (front & back).
4713	R1542	03/11/95	4	Road Series.
4738-49	A Series	03/05/96		HiFi cab assy. Black, rosewood, walnut, cherry.
5105-07	QX Series	06/08/98	8	Sound reinforcement cabinets.
5113	G12H-80		8	Marshall Wolverine.
5114-15	G12T-100		4, 8	Special for Fender.
5138	Rockdriver Pro 100	23/6/99	8	Hughes & Kettner.
5143	Vox Blue 8	15/7/99	8	Custom made for Vox (Korg).
5144-47	CXi Range	29/7/99	8	PA cab assy. Black. Birch plywood.
5166-68	Frontline 12	15/12/99	8	Cast frame LF speakers 12in to 18in.
5257, 72	G12 Century	19/1/00	8, 16	Neodymium magnet guitar speaker.
5174-78	Truvox	19/1/00	8	Pressed steel LF speaker 10in - 15in.
5260	NTi-1550	25/4/01	8	15in cast frame LF speaker, inverted neodymium magnet.
5343, 44	CDX1-1415/1425	07/02/02		Neo mag. Assy. Front firing.
5347, 48	Century Vintage		8, 16	Neo mag assy. Vintage cone.
5363	CDX1-1745	24/07/03		Was 1750.
5367	CDX1-1430	10/12/03	8	Neo magnet, front firing.
5373, 81	G10 Vintage		8, 16	Vintage 30 cone assy.
5387-91	FTR Range	29/03/04		Cast frame PA speaker 12in to 18in.
5471, 72	G12 Gold		8, 15	Alnico magnet. New V/C.

T Number	Model	Date Issued	Ω	Description
5485	CDX1-1730	13/04/06	8	T5363 with neo mag assy.
5600	G12-EVH		15	As T1221-67. Black can & bespoke label, keep solder tags.
5602	FTR18-4080HDX	15/09/06	8	4in V/C. Longer throw.
5612-5615	TN Series	07/12/06	8	Pressed steel neo magnet mid/bass speakers
5619	BN12-300S	11/12/06	8	Standard BN range (similar to T5445)
5640	CDX14-3050	07/03/07	8	Deep drawn Ti diaphragm.
5646, 47	G10 Greenback		8, 16	1.75in V/C. New cone.
5658, 70	G12-EVH	20/09/07	8, 15	As T5600 with faston connector.
5671, 82	G10 Gold		8, 15	Alnico magnet. New V/C.
5704	NTR06-1705D	6/25/08	16	Neo mag. assy. Cast frame mid/bass.
5797	G12-50GL		8	Special for Randall. G12M cone assy. Modified V/C. Lynchback.
5801	AN3510 8ohm	12/6/10	8	Full range. Steerable line array speaker.
5806	CF18VJD 8ohm	12/17/10	8	Ferrite, 5in V/C 1600W rms.
5848	CDX1-1747 8ohm	9/30/11	8	As T5343 with polyimide dia assy
5863	CF1840JD	1/10/12	8	4in V/C.
T5864, 71	G12M-65		8, 16	Creamback.
5890, 91	G12H-75 Creamback	11/8/12	8, 16	As T5864, 50oz magnet.
5910	FTX0820	7/3/13	8	Common motor coaxial.
5911	FTX1225	7/3/13	8	Common motor coaxial.
5924, 29	G12-35XC Ltd Edition	11/7/13	8	42 oz magnet.
5925, 30	A Type	11/28/13	8, 16	35 oz magnet. New cone.
5948	G15V-100 Fullback	9/29/14	8	50 oz magnet. 2in V/C.
5949	CF0820BMB	10/3/14	8	Bass/ Mid.bass. Rubber surround.
5953, 54	Celestion Cream	11/28/14	8, 16	90W. Alnico mag assy.
5968-70	Pulse	4/17/15	8	Retail branded bass guitar speakers 10in, 12in, 15in.
5977, 81	Neo Creamback	7/14/15	8	New style neodymium magnet assembly. Creamback cone assy.
5984	Axi2050	11/2/15	8	Axiperiodic Ti diaphragm, 5in V/C. Wideband compression driver
5995	FTX1530	2/22/16		Common motor coaxial.
6317	CN0515M	9/6/16	16	Similar to T5966 / S7652.
6328, 29	G12H-150 Redback	11/7/16	8, 16	2in V/C. Creamback cone.
6350	CDX1-1732	8/21/17	8	As T5485 CDX1-1730 but polyimide dia assy. New cover.
6351	F12-X200 8ohm	8/21/17	8	CDX1-1425 on rear of magnet. X-over inc. New can.
6370	Neo 250 Copperback	1/16/18	8	Curved cone, CE. 2.5" V/C.
6998	TSQ1845MD	1/12/21	8	New style LF loudspeaker. 4.5in V/C. Longer throw.
6472	G12M-50 Hempback	8/14/20	8	35oz Magnet. New hemp cone.
6605	PowerProX18	5/5/22	8	4in V/C.

Appendix II: Patents Granted

Title	Description
Improvements in telephone receivers and like instruments	Centrally supported light diaphragm of paper with cane reinforcement strips for piston behaviour.
Improvements in or relating to wireless apparatus	Cabinet with a hinged cover under which the speaker is mounted.
Improvements in or relating to gramophones	Stylus attached to articulated arm and directly connected to a radiating diaphragm.
Improvements in or relating to diaphragms for telephone instruments	Paper diaphragm stiffened by cane but clamped at the edge to increase radiating area and add damping to increase bandwith.
Improvements in or relating to telephone receivers	Flat diaphragm for a telephone receiver, attached to magnet by cane ribs.
Improvements in or relating to diaphragms for telephones, talking machines and the like	Edge attached to a corrugated cylinder.
Improvements in or connected with diaphragms for acoustic instruments	Ridged and grooved diaphragm to encourage piston behaviour.
Improvements in and relating to loud speakers and the like instruments	Cylindrical edge attached to rocking mass allowing diaphragm to self centre under temperature or positional change, providing vibration isolation against transit shocks.
Improvements in or relating to acoustic devices	Armature mounted in the concave face, amplifying bowl at the rear.
Improvements in and relating to sound reproducing and receiving instruments	Maintains radiating area at L.F. while allowing reduced displacement in armature and smaller, stiffer radiators.
Improvements in electromagnetic apparatus for use in the reception or reproduction of sound	Multiple diaphragms driven by one magnet.
Improvements in or relating to loud speaking apparatus for wireless or other purpose	Collapsible baffles for easy transportation.
Improvements in or relating to sound reproducing devices	Multiple diaphragms of different radiating area driven by one magnet.
Sound reproducer for gramophones	Travelling diaphragm attached to stylus.
Loud speaker and like acoustic apparatus	Flat, spiral springs to centre the diaphragm.
Improvements relating to telephone instruments	Spider and/or surround supported on soft material to reduce fragility.
Improvements relating to telephone instruments and like electro-magnetic devices	Variable armature cross-section allows wider bandwidth and damping for smoother response (a kind of 'electromagnetic crossover').
Improvements relating to telephone instruments	Geared, interconnected adjustment for multiple armatures.
Waxwork, robot, and like figures	(2-way) intercom system for public spaces, uses a human caricature or robot.
Improvements in electro-magnetic vibratory devices such, for example, as loudspeakers	Inclined armature reduces resonance frequency and distortion.

Publication nos.	Date Granted	Inventor(s)	Applicant(s)
GB230552(A)	16 March, 1925	Eric Mackintosh	Eric Mackintosh
GB238366(A)	20 August, 1925	Cyril French; Ralph French	Cyril French; Ralph French
GB245547(A)	14 January, 1929	Ralph French; Harriet French	Celestion Radio Co. Ltd
GB245704(A)	14 January, 1926	Eric Mackintosh; Cyril French	Celestion Radio Co. Ltd
GB262255(A)	09 December, 1926	Edgar French	Celestion Radio Co. Ltd
GB266513(A)	03 March, 1927	Ralph French	Celestion Radio Co. Ltd
GB291526(A)	05 June, 1928	Ralph French	Celestion Radio Co. Ltd
GB293200(A)	05 July, 1928	Eric Mackintosh; Cyril French	Celestion Ltd
GB301559(A)	06 December, 1928	Cyril French; Edward J Marriott	Celestion Ltd
GB306263(A)	21 February, 1929	Eric Mackintosh; Cyril French	Celestion Radio Company
GB304487(A)	24 January, 1929	Cyril French	Celestion Radio Company
GB306278 (A)	21 February, 1929	Cyril French; Edward J Marriott	Celestion Radio Company
GB318766(A)	12 September, 1929	Cyril French	Celestion Radio Company
GB320229(A)	10 October, 1929	Eric Mackintosh	Celestion Radio Company
GB321806(A)	21 November, 1929	Cyril French	Celestion Radio Company
GB325976 (A)	06 March, 1930	Cyril French	Celestion Ltd
GB330547 (A)	10 June, 1930	Cyril French	Celestion Ltd
GB331336 (A)	03 July, 1930	Cyril French	Celestion Ltd
US1774653	02 September, 1930	Cyril French; Edward J Marriott	Celestion Ltd
GB341647 (A)	22 January, 1931	Eric V. Mackintosh	Celestion Ltd

Title	Description
Improvements relating to acoustic instruments	Convoluted baffles, having the effect of fitting larger baffles into smaller spaces.
Improvements relating to sound reproducing and like instruments	Quick-release stylus holder (complementary to GB356353).
Improvements relating to electro-magnetic vibratory devices particularly for use in the recording and reproduction of sound	Improved armature for stylus (complementary to GB355880).
Improvements in or relating to the manufacture of permanent-magnet electrical apparatus	Assembling a permanent magnet in a demagnetised state, cleaning of unwanted particles, followed by permanent magnetisation.
Improvements in and relating to diaphragms for sound reproducing instruments	Moulded pulp diaphragm impregnated with resin.
Improvements in or relating to protective covers for apertures for loudspeaking instruments, microphones, and other articles	Water-tight cover with non-rigid surround.
Improvements in or relating to electrodynamic loudspeakers or microphones	Centring mechanism for magnet assembly.
Improvements in or relating to gramophone pick-ups and the like	Improved suspension and damping for gramophone pickup.
Improvements in or relating to gramophone pick-ups and the like	Improved ribbon-type gramophone pickup.
Electromagnetic transducer	Shallow drive unit due to magnet mounted in concave face of cone.
Improvements in permanent magnet assemblies for moving coil loudspeakers, microphones and other instruments	Non-magnetic sleeve reduces machining operations for magnet assembly.
Improvements in or relating to loud speakers and like electro-acoustic transducers	A simplified construction method for re-entrant horns.
Improvements in or relating to moving coil loudspeakers and microphones	Improved method of locating and securing a loudspeaker's magnet yoke.
Improvements in or relating to electro-acoustic transducers	Improved and simplified transducer construction particularly adapted to manufacture in very small sizes.
Improvements in or relating to electro-acoustic transducers	Shaping outer edge of near conical diaphragm to achieve more accurate alignment when fixing to chassis.
Improvements in or relating to electronic apparatus incorporating electrostatic loudspeakers	Improved structural arrangement of two electrostatic transducers for stereophonic reproduction.
Improvements in or relating to electrostatic loudspeakers	Mounting electrodes on the diaphragms of electrostatic transducers
Improvements in or relating to magnetic structures including a permanent magnet	Improved method of welding together a magnet with conductive material when making a permanent magnet assembly.
Improvements in or relating to sound reproducing apparatus	A crossover circuit enabling the combination of an electrostatic loudspeaker with a moving coil loudspeaker.
Improvements in or relating to electro-mechanical transducers	A simple and rapid method of constructing small transducers.

Publication nos.	Date Granted	Inventor(s)	Applicant(s)
GB337264 (A)	30 October, 1930	Cyril French	Celestion Ltd
GB355880 (A)	03 September, 1931	S. James Tyrrell	Celestion Ltd
GB356353 (A)	10 September, 1931	S. James Tyrrell	Celestion Ltd
GB410491 (A)	16 May, 1934	Henry S. Tenny	British Rola Company Ltd
GB485550 (A)	18 May, 1938	Josiah A. Briscoe; James H. Nelson	Celestion Ltd; Murphy Radio Ltd
GB552676 (A)	20 April, 1943	Dennis H. Marlow	British Rola Ltd
GB595739 (A)	15 December, 1947	Leslie R. Ward	Truvox Engineering Co Ltd
GB611184 (A)	26 October, 1948	Frederick Clark	Truvox Engineering Co Ltd
GB622241 (A)	28 April, 1949	Frederick Clark	Truvox Engineering Co Ltd
US2537723 (A)	09 January, 1951	Leslie R. Ward	Truvox Engineering Co Ltd
GB674875 (A)	02 July, 1952	S. James Tyrell; Arthur C. Young	Rola Celestion Ltd
GB742889 (A)	04 January, 1956	Leslie R. Ward; S. James Tyrrell	Rola Celestion Ltd
GB801695 (A)	17 September, 1958	S. James Tyrrell; Arthur C. Young	Rola Celestion Ltd
GB809481 (A)	25 February, 1959	Arthur C. Young	Rola Celestion Ltd
GB889967 (A)	21 February, 1962	Arthur C. Young	Rola Celestion Ltd
GB894900 (A)	26 April, 1962	S. James Tyrrell; Arthur C. Young	Rola Celestion Ltd
GB884364 (A)	13 December, 1961	Arthur C. Young	Rola Celestion Ltd
GB858977 (A)	18 January, 1961	Arthur C. Young	Rola Celestion Ltd
GB842836 (A)	27 July, 1960	Arthur C. Young	Rola Celestion Ltd
GB885991 (A)	03 January, 1962	Arthur C. Young	Rola Celestion Ltd

Title	Description
Improvements in or relating to electro-mechanical transducers	Improved form of loudspeaker construction particularly suitable for producing a device of small physical size (e.g. for transistor radio receiver).
Improvements in or relating to moving coil loudspeakers and like transducers	Improved method of manufacturing a permanent magnet assembly with a U shaped yoke.
Improvements in or relating to electro-acoustic transducers	Improved form of magnet assembly construction used with small diaphragm loudspeakers.
Improvements in or relating to permanent magnet moving coil loudspeakers	Use of a stepped diameter yoke to simplify magnet assembly construction.
Improvements in or relating to electro-acoustic transducers of the moving coil permanent magnet type	A more efficient way of constructing a magnet assembly, preventing adhesive from getting in flux gap.
Improvements in or relating to electro-acoustic transducers	Improved method of constructing a loudspeaker magnet assembly using an annular ceramic ring magnet.
Magnet assemblies for moving coil electro-acoustic transducers	A loudspeaker magnet assembly that allows for greater voice coil movement reducing the likelihood of damage in high power transducers.
Loudspeaker Diaphragm	Midrange compression driver diaphragm able to handle high power input.
Ribbon transducers.	To provide an improved (damped) end mounting for a ribbon transducer to reduce possibility of ribbon fracture.
Ribbon transducers.	A simple means of suspending and centering the ribbon of a ribbon transducer, capable of easy adjustment.
Compression driver diaphragm clamping means	"Sound Castle" compression driver diaphragm clamping.
Acoustic horns for loudspeakers incorporate vibration damping material	Aluminium horn with inegrated damping: "No Bell".
Magnet system for loudspeakers	Using two magnets separated by a plate to construct a loudspeaker motor that is lighter, more compact and more efficient.
Loudspeaker assembly	Front mounted neodymium magnet used on NTi drivers.
Loudspeaker pole piece and loudspeaker assembly	Forging heat sinking fins into a loudspeaker magnet pole piece.
Loudspeaker chassis formed from segments	Casting large loudspeaker chassis with a tool that forms a quarter of the entire chassis.
Electro Acoustic Diaphragm	Axiperiodically shaped diaphragm for use with wideband compression drivers.
Acoustic phase plug	Unique phase plug design with circumferential undulations for use with "Axiperiodic" diaphragm.
Acoustic waveguides	Routing sound waves in specific directions without disrupting coherence of wavefront (lensguide).

Publication nos.	Date Granted	Inventor(s)	Applicant(s)
GB886757 (A)	10 January, 1962	Arthur C. Young	Rola Celestion Ltd
GB943968 (A)	11 December, 1963	Arthur C. Young	Rola Celestion Ltd
GB991945 (A)	12 May, 1965	Arthur C. Young	Rola Celestion Ltd
GB1018165 (A)	26 January, 1966	Arthur C. Young	Rola Celestion Ltd
GB1097974 (A)	03 January, 1968	Arthur C. Young	Rola Celestion Ltd
GB1136597 (A)	11 December, 1968	Arthur C. Young	Rola Celestion Ltd
GB1321581 (A)	27 June, 1973	Leslie R. Ward	Rola Celestion Ltd
US3780232 (A)	18 December, 1973	Leslie R. Ward	Rola Celestion Ltd
EP0404487 (A2); EP0404487 (A3); EP0404487 (B1)	19 April, 1995	Graham Bank; Harold Charles Pinfold	Celestion Int Ltd
EP0404488 (B1)	24 August, 1994	Graham Bank; Harold Charles Pinfold	Celestion Int Ltd
GB2322503 (A); GB2322503 (B)	01 November, 2000	Mark A. Dodd	Celestion Int Ltd
GB2325603 (A); GB2325603 (B)	01 November, 2000	Mark A. Dodd	Celestion Int Ltd
US2002094107 (A1); US6563932 (B2)	24 April, 2003	Paul Cork	KH Technology Corp
US2002106101 (A1); US7016514 (B2)	21 March, 2006	Ian S. White; Richard C. Klein	KH Technology Corp
US2002094105 (A1)	11 August, 2004	Duncan Boniface	KH Technology Corp
GB2458689 (A); GB2458689 (B)	23 May, 2012	Ian S. White; Paul V. Cork	GP Acoustics Uk Ltd
US9467782 (B2)	11 October, 2016	Jack A, Oclee-Brown; Mark A. Dodd	GP Acoustics Uk Ltd
ES2855033 (T3)	23 December, 2020	Jack A, Oclee-Brown; Mark A. Dodd	GP Acoustics Uk Ltd
US11626098 (B2); US2021110808 (A1)	11 April, 2023	Jack A, Oclee-Brown; Mark A. Dodd	GP Acoustics Intl Ltd

Appendix III: Managing Directors/General Managers

Year	Managing Director	Year	Managing Director
1924	Cyril French	1949	Jim Tyrrell
1925	Cyril French	1950	Jim Tyrrell
1926	Cyril French	1951	Jim Tyrrell
1927	Cyril French	1952	Jim Tyrrell/Billy Page
1928	Cyril French	1953	Jim Tyrrell/Billy Page
1929	Cyril French	1954	Jim Tyrrell/Billy Page
1930	Cyril French	1955	Jim Tyrrell/Billy Page
1931	Cyril French	1956	Jim Tyrrell/Billy Page
1932	Cyril French	1957	Jim Tyrrell/Billy Page
1933	Cyril French	1958	Jim Tyrrell/Billy Page
1934	Cyril French	1959	Jim Tyrrell/Billy Page
1935	R. B. Page	1960	Jim Tyrrell/Billy Page
1936	R. B. Page	1961	Jim Tyrrell/Billy Page
1937	Stephen de László	1962	Jim Tyrrell/Billy Page
1938	Stephen de László	1963	Jim Tyrrell/Billy Page
1939	Patrick de László	1964	Jim Tyrrell/Billy Page
1940	Patrick de László	1965	Jim Tyrrell/Billy Page
1941	Patrick de László	1966	Jim Tyrrell/Billy Page
1942	Patrick de László	1967	Neil MacKinlay
1943	Patrick de László	1968	Neil MacKinlay
1944	Patrick de László	1969	Neil MacKinlay
1945	Patrick de László	1970	Neil MacKinlay
1946	Patrick de László	1971	Neil MacKinlay
1947	Jim Tyrrell	1972	Neil MacKinlay
1948	Jim Tyrrell	1973	Neil MacKinlay

Year	MD/General Manager	Year	MD/General Manager
1974	John Church	1999	Richard Wear
1975	John Church	2000	Frank DiGirolamo
1976	John Church/Stewart Black	2001	Frank DiGirolamo
1977	Stewart Black/Colin Aldridge	2002	Frank DiGirolamo
1978	Colin Aldridge	2003	Frank DiGirolamo
1979	Colin Aldridge	2004	Brian Li/Nigel Wood
1980	Colin Aldridge	2005	Nigel Wood
1981	Colin Aldridge	2006	Nigel Wood
1982	Colin Aldridge	2007	Nigel Wood
1983	Colin Aldridge/David Marsh	2008	Nigel Wood
1984	David Marsh	2009	Nigel Wood
1985	David Marsh	2010	Nigel Wood
1986	David Marsh	2011	Nigel Wood
1987	Gordon Provan	2012	Nigel Wood
1988	Gordon Provan	2013	Nigel Wood
1989	Gordon Provan	2014	Nigel Wood
1990	Gordon Provan	2015	Nigel Wood
1991	Gordon Provan	2016	Nigel Wood
1992	Gordon Provan	2017	Nigel Wood
1993	Gordon Provan	2018	Nigel Wood
1994	Gordon Provan	2019	Nigel Wood
1995	Gordon Provan	2020	Nigel Wood
1996	Andrew Osmond	2021	Nigel Wood
1997	Richard Wear	2022	Nigel Wood
1998	Richard Wear	2023	Nigel Wood

Appendix IV: Celestion Date Codes

Most Celestion chassis drivers have been marked with a date code (two numbers and two letters), denoting the exact date of manufacture. These codes are added on the production line and are found in different locations depending on the age (or type) of speaker. They can be:

1. Stamped on the rim of the chassis (pre-1952);

2. stamped on the front gasket (1952-1967 and modern alnico guitar speakers);

3. on the speaker's chassis leg (1968-2001 and Heritage Series guitar speakers);

4. or printed onto a white ID label stuck onto the magnet's edge (2001 to date).

If you can track down the date code, then all you need to do is refer to the following table, which provides a record of the date codes used throughout the years, and you should be able to decipher the code.

Date codes are expressed in the form of two digits (representing the day of the month 01 to 31) and two letters, one being the month, the other being the year. You'll see from the table below that the format varies from time to time, sometimes the two digits denoting the day come first (DDMY), other times they come at the end (MYDD). Be mindful of the variation when working out your speaker's age.

From the speaker in the photo below we can see that the date code is BL14, so the format is MYDD. That makes the date for this speaker February 14, 1978.

Month	1944–1955 (DDMY)	1956–1968 (DDMY)	1969–1990 (MYDD)	1991–2014 (DDMY)	2015–2024 (MYDD)
A – Jan	A – 1944	A – 1956	B – 1969	A – 1991	A – 2015
B – Feb	B – 1945	B – 1957	C – 1970	B – 1992	B – 2016
C – Mar	C – 1946	C – 1958	D – 1971	C – 1993	C – 2017
D – Apr	D – 1947	D – 1959	E – 1972	D – 1994	D – 2018
E – May	E – 1948	E – 1960	F – 1973	E – 1995	E – 2019
F – Jun	F – 1949	F – 1961	G – 1974	F – 1996	F – 2020
G – Jul	G – 1950	G – 1962	H – 1975	G – 1997	G – 2021
H – Aug	H – 1951	H – 1963	I/J – 1976	H – 1998	H – 2022
I/J – Sep	I/J – 1952	I/J – 1964	K – 1977	J – 1999	J – 2023
K – Oct	K – 1953	K – 1965	L – 1978	K – 2000	K – 2024
L – Nov	L – 1954	L – 1966	M – 1979	L – 2001	
M – Dec	M – 1955	M – 1967	N – 1980	M – 2002	
		A – 1968	O – 1981	N – 2003	
			P – 1982	P – 2004	
			Q – 1983	Q – 2005	
			R – 1984	R – 2006	
			S – 1985	S – 2007	
			T – 1986	T – 2008	
			U – 1987	U – 2009	
			V – 1988	V – 2010	
			W – 1989	W – 2011	
			X – 1990	X – 2012	
				Y – 2013	
				Z – 2014	

With acknowledgement to Michael Doyle who did some great research work in the early 1990s verifying the format of pre-1968 date codes. You can read more about that in his fine book, *The History of Marshall*. More recently, Brian Harding has done much in-depth work investigating and compiling date codes from the early 1950s, going as far back as 1944: we're grateful to him for that information and happy to include it here. You can read more about Brian's investigations on his excellent website, bygonetones.com

APPENDIX IV: CELESTION DATE CODES

Acknowledgements

I'd like to begin by thanking Jerry Gilbert, whose significant experience in music journalism, knowledge of the early days of the sound reinforcement industry and unerring ability to locate interviewees and conduct those interviews with persistence and diplomacy has been invaluable. I gratefully acknowledge and thank Jerry for the writing work he's done and his assistance in bringing a logical order to the book, which I hope makes the whole thing readable as a complete story.

Thanks must also go to Terry Marshall, whose additional insights on the early days of Marshall Amplification were gold dust. Gratitude is also extended to music industry titans Lyndon Laney and Cliff Cooper (Orange Amplification) for letting us into the origin stories of their legendary companies. I'm likewise very grateful to our good friends and partners in sound reinforcement, in particular Pat Quilter (Quilter Labs, QSC), Tony Andrews (Funktion One) as well as Dominic Harter and Phil Anthony (Martin Audio) and their former team member, turned pro audio consultant Jason Baird.

Additionally Chris Hewitt (CH Vintage Audio) was generous with images, information and expertise on early Vox amps and WEM PA in particular, as well as the PA industry of the early 1970s. His knowledge is fascinating and extensive.

Special mention should also go to employees and partners of Celestion past and present who have contributed with their memories and stories, alphabetically: Andy Farrow, Brian Coviello, Bob Smith, Dee Potter, Ed Form, Ian White, Jack Kelly, Jasmine Booth, Julian Wright, Mark Dodd, Nigel Wood, Paul Airey, Paul Cork, Peter Ambrose, Peter Wellikoff, Richard Klein, Richard Wear and Sonia Jackson. Not to mention Rick Wakeman and Robbie Gladwell!

This book carries on a noble tradition, documenting the many faceted history of Celestion. It stands on the shoulders of great work done by Julian Wright who spent several years during his time at the company compiling the original *History of Celestion* book, first published in 1995. His book in turn was developed from an earlier history assembled by Gordon Kinsey more than a decade earlier, to which Julian added much important information, particularly relating to the very early days of the company. Julian's skills in moving through genealogy and other history-based websites has uncovered some truly fascinating material for this book too, enabling us to paint a much more detailed picture of the genesis of the company.

In editing *A Century of Sound* I've brought together a great deal of additional research, much gleaned from contemporary newspaper reporting and material from trade magazines, together with information gathered from face-to-face interviews. Other published works have been referenced, most notably Michael Doyle's excellent *History of Marshall* and Jim Elyea's *Vox Amplifiers: The JMI Years*. I've also relied on other wellsprings of online information and images, namely Brian Harding at Bygone Tones, Shem Kaczynski at Audio Nostalgia, Nick Orchard at voxac30.org as well as Kevin Fulcher's detailed Ditton Works YouTube channel and associated Facebook group. Brian and Kevin were also kind enough to help with some fact checking.

Finally, much use was made of the existing Celestion archive with its collection of old brochures, advertising, articles and photography, newsletters and even inter-departmental memos much of which has become the glue holding this story together. The result is a more complete story of some of the key events, products and personalities that have helped forge this company, which is truly a legend of the British audio industry. I've had to make educated guesses a time or two and very occasionally there is a little speculation, only to keep the story moving in the right direction, you understand.

JP, November 2023

Picture credits

Alamy: p7, p10, p11, p25, p62, p67, p68, p80, p92, p120; *Beat International*: p60; Brian Coviello: p156; Brian Harding: p129, p130; Bridgeman Images: p38; Chad Lee: p194; Chris Hewitt (CH Vintage Audio): p72, p79; Kevin Fulcher: p111, p119, p150; Laney Amplification: p75; © Michael Mesker: p194 *tr*; Mike Arbon/Rollerbury: p134; Nick Orchard: p58; OMEC: p77; Science Photo Library: p39; Shem Kaczynski: p86, p110; Tyler Chandler de Faye: p15 *c*.

Published 2023

Copyright © 2023 by Celestion.

All rights reserved. No part of this book may be reproduced or transmitted in any form or by any means, electronic or mechanical, including photocopying, recording, or by any information storage and retrieval system without the written permission of Celestion, except where permitted by law.

ISBN: 978-1-3999-7336-6

www.celestion.com

Project managed by John Paice
Cover and interior design by Paul Palmer-Edwards
Written by John Paice and Jerry Gilbert
Printed and bound by CPI Group (UK) Ltd

The PowerProX18, 18" cast chassis low frequency loudspeaker.